Cybersecurity Strategies and Best Practices

A comprehensive guide to mastering enterprise
cyber defense tactics and techniques

Milad Aslaner

Cybersecurity Strategies and Best Practices

Group Product Manager: Pavan Ramchandani

Publishing Product Manager: Khushboo Samkaria

Book Project Manager: Uma Devi

Senior Editor: Divya Vijayan

Technical Editor: Irfa Ansari

Copy Editor: Safis Editing

Proofreader: Divya Vijayan

Indexer: Tejal Daruwale Soni

Production Designer: Ponraj Dhandapani

Senior DevRel Marketing Executive: Linda Pearlson

DevRel Marketing Coordinator: Marylou De Mello

First published: May 2024

Production reference: 1240424

Published by Packt Publishing Ltd.
Grosvenor House
11 St Paul's Square
Birmingham
B3 1RB, UK.

ISBN 978-1-80323-005-4

www.packtpub.com

This book is dedicated to my wife, Salpie, my life partner, for her relentless support; to our son, Raphael, my best friend and source of motivation; to Balou, our sweet bundle of joy; and to my siblings, Aydin and Aylin, for shaping me through their unwavering presence in our family's journey.

– Milad Aslaner

Contributors

About the author

Milad Aslaner is a cybersecurity thought leader with over two decades of experience in the field, specializing in security architecture, security operations, and incident response. With a career spanning multiple domains within cybersecurity, Milad has established himself as an expert in the industry. Beginning his journey in cybersecurity consultancy, Milad honed his expertise in solving multifaceted security challenges, laying the foundation for his illustrious career. His experience encompasses a wide array of roles, including leadership positions, where he has consistently demonstrated his prowess in navigating complex cybersecurity landscapes. As a published author and recognized authority in the cybersecurity community, Milad frequently shares his insights and knowledge through speaking engagements at conferences and panels. His contributions to the field have been instrumental in shaping the discourse around cybersecurity best practices and emerging trends.

"Learning is not attained by chance, it must be sought for with ardor and attended to with diligence" – Abigail Adams.

I extend my deepest gratitude to all who have supported me in writing and publishing this book. Your contributions, encouragement, and insights have been invaluable on this journey of continuous learning. Thank you for your unwavering commitment and dedication.

About the reviewers

Sina Manavi, is a seasoned cybersecurity expert with over 17 years of experience in global information security leadership and strategic roles across diverse industries, including consulting, banking, insurance, and logistics. His extensive expertise encompasses Multi-Cloud environments (Azure, Google, and Oracle) as well as on-premises setups, where he has managed security services, product oversight, and led various security domains and teams.

Holding an array of advanced certifications—ISO 27001, C|CISO, CISM, CISA, CDPSE, CEH, and CHFI—Sina exemplifies the pinnacle of professional qualification in his field. His scholarly contributions are showcased across his YouTube Channel, LinkedIn, and Google Scholar profiles. Moreover, he has lent his expertise as a technical reviewer for numerous cybersecurity books, including *Kali Linux Wireless Penetration Testing Essentials* and *Threat Hunting in the Cloud: Defending AWS, Azure and Other Cloud Platforms Against Cyberattacks*.

António Vasconcelos, a tech maven with 23 years in the IT industry, specializes in cybersecurity. His journey includes a decade at Microsoft, where he notably served as a product manager for EDR and XDR. António's expertise was further honed at SentinelOne, embracing roles such as field CISO and product manager for XDR. Presently, he is making strides at Zero Networks. A firm advocate for robust cyber defense, António excels in merging business acumen with cybersecurity, ensuring organizations navigate the digital realm securely and efficiently.

Josh Mason, the director of cyber training and vCISO at Arbitr, brings a wealth of expertise as a former pilot and cyber warfare officer in the United States Air Force. With a focus on building cyber programs and developing training, Josh is a key contributor to the field of cybersecurity.

As a technical editor for this book, Josh leverages his extensive background to provide valuable insights and guidance. His military experience, including building training programs and teaching at renowned institutions, such as the US Air Force Special Operations School and the DoD Cyber Crime Center's Cyber Training Academy, uniquely positions him to offer practical and strategic perspectives.

Table of Contents

Preface   xiii

1

Profiling Cyber Adversaries and Their Tactics   1

Types of threat actors   1
Summary of threat actor categories   3

Motivations and objectives of threat
actors   9

Tactics, Techniques, and Procedures
(TTPs)   10

Real-world examples of cyberattacks
and consequences   13

Nation-state actors: NotPetya attack (2017)   13
Nation-state actors: SolarWinds supply chain
attack   14
Cybercriminals: WannaCry ransomware
attack (2017)   16
Cybercriminals: Colonial Pipeline
ransomware attack   17

Summary   19

2

Identifying and Assessing Organizational Weaknesses   21

Understanding organizational
weaknesses and vulnerabilities   22

Types of organizational weaknesses   22
Types of organizational vulnerabilities   23
Real-world examples   24
Techniques for identifying and assessing
weaknesses   25
Security audits   25
Vulnerability assessments   26
Threat modeling   27
Penetration testing   29

Social engineering tests   30

Conducting risk assessments   32
Risk assessment methodologies   32
Identifying assets and establishing the scope   33

Prioritizing risks and developing
mitigation strategies   36
Documentation and reporting   37
Monitoring and reviewing   38
Prioritizing and remediating weaknesses   39
Understanding risk and impact levels   39

Risk mitigation strategies   40   Continuous monitoring and reassessment   42

Attack surface reduction   41   **Summary**   **43**

3

Staying Ahead: Monitoring Emerging Threats and Trends   45

The importance of monitoring
emerging threats and trends   46
Understanding the cybersecurity landscape   46
The risks of emerging threats   48
The role of threat intelligence   49
From awareness to action   50

The attacker's mindset   52
The significance of understanding the
attacker's perspective   52
Motivations and objectives of attackers   53

Psychological and behavioral traits of attackers   54
The role of the attacker's mindset in
strengthening cybersecurity   55
Ethical considerations and legal boundaries   56
Ethical hacking and responsible disclosure   57

The role of innovation in cybersecurity 58
The benefits of and need for innovation   58
Driving innovation within organizations   59
Emerging technologies and future trends   60

Summary   **61**

4

Assessing Your Organization's Security Posture   63

The components of a comprehensive
security posture   63
Evaluating security technologies   64
Understanding the role of security processes   65
The human factor in a security posture   66

Effective metrics for security
programs and teams   67
Understanding the importance
of security metrics   67
Selecting the right metrics   68
Implementing and tracking security metrics   71

Asset inventory management and its
role in security posture   71
Understanding asset inventory in cybersecurity   71
Building a comprehensive asset inventory   72
Maintaining and updating asset inventory   73
Continuously monitoring and improving
your security posture   74
Implementing continuous monitoring practices   74
Responding to incidents and implementing
remediation measures   75
The technological landscape in security posture   76

Summary   **78**

5

Developing a Comprehensive Modern Cybersecurity Strategy   79

Key elements of a successful cybersecurity strategy   80

Foundational principles and components   80

Setting objectives and goals   81

The role and significance of each element   82

Aligning cybersecurity strategy with business objectives   83

Correlation of organizational goals and cybersecurity endeavors   83

Prioritizing cybersecurity based on business impact   84

Communicating cybersecurity's value to stakeholders   86

Risk management and cybersecurity strategy   87

Integrating risk management methodologies in strategy formulation   87

Conducting comprehensive risk assessments   89

Prioritization of mitigation strategies   91

Incident response planning and preparedness   92

Designing tailored incident response procedures   93

The incident management life cycle   95

Tools, technologies, and human elements in incident response   96

Security awareness and training programs   97

Tailored training for organizational roles   98

Continuous evaluation and improvement   99

Fostering a security-first mindset   101

Summary   102

6

Aligning Security Measures with Business Objectives   103

The importance of aligning security with business objectives   104

The critical role of cybersecurity in business environments   104

Connecting business objectives and security measures successfully   105

Measuring the impact and value of aligned cybersecurity initiatives   107

Prioritizing security initiatives based on risk and business impact   109

The importance of risk assessment and BIA   109

Prioritizing security initiatives with frameworks   110

Communicating prioritized security initiatives   112

Communicating the value of security investments   113

Translating technical metrics to business value   113

Developing effective communication strategies   115

Engaging and building trust with stakeholders   117

Summary   118

7

Demystifying Technology and Vendor Claims   119

Understanding technology and vendor claims   120

Deciphering the language of cybersecurity claims   120

Separating facts from marketing in vendor claims   123

Evaluating the substance of cybersecurity solutions   124

Critically analyzing claims   125

Developing a skeptical mindset   126

Contextual analysis of vendor claims   127

Identifying biases and unsupported assertions   128

Utilizing analyst and third-party testing reports   130

Understanding and accessing external resources with practical examples   130

Interpreting methodologies and results   132

Applying findings to an organizational context   133

Thoroughly assessing vendors   134

Evaluating vendor credibility and track record   135

Analyzing customer feedback and post-sale support   136

Aligning vendor offerings with organizational requirements   137

Summary   138

8

Leveraging Existing Tools for Enhanced Security   139

Identifying existing and required tools and technologies   139

Cataloging your cybersecurity arsenal   140

Assessing tool effectiveness and relevance   141

Identifying gaps and future needs   142

Repurposing and integrating tools for enhanced security   143

Repurposing of cybersecurity tools   143

Integration of security tools   144

Maximizing efficiency through tool synergy   145

Optimizing tool usage for maximum value   146

Advanced configuration and customization of tools   146

Performance monitoring and regular audits   147

Training and knowledge sharing   148

Summary   149

9

Selecting and Implementing the Right Cybersecurity Solutions   151

Factors to consider when selecting cybersecurity solutions   152

Understanding the threat landscape   153

Assessing system compatibility and integration   154

Scalability and future-proofing cybersecurity
solutions   159

Compliance and industry standards in
cybersecurity solutions   160

**Best practices for selecting security
tools   162**

Conducting comprehensive market research   163

Involving key stakeholders in the selection
process   164

Performing risk assessment and management   166

Evaluating cost-effectiveness and ROI in
cybersecurity solutions   167

**Implementing and integrating
cybersecurity solutions   169**

Developing a strategic implementation plan
for cybersecurity solutions   170

User training and adoption in cybersecurity
implementation   172

Monitoring, maintaining, and regularly
updating cybersecurity solutions   174

Summary   176

10

Bridging the Gap between Technical and Non-Technical Stakeholders   177

**The Importance of Effective
Communication and Collaboration   178**

Understanding communication barriers in
cybersecurity   178

The role of effective communication in
cybersecurity success   179

**Strategies for successful collaboration
between technical and non-technical
stakeholders   180**

**Translating technical concepts for
non-technical stakeholders   181**

Simplifying complex cybersecurity terminology 182

Contextualizing cybersecurity in business
terms   183

Effective visualization and presentation of
cybersecurity data   184

Strategies for successful collaboration   185

Building cross-functional cybersecurity teams 186

Establishing regular cybersecurity workshops
and training sessions   187

Implementing collaborative cybersecurity
decision-making processes   188

Summary   189

11

Building a Cybersecurity-Aware Organizational Culture   191

**The importance of a cybersecurity-
aware organizational culture   192**

Understanding cybersecurity as a business
imperative   192

Assessing the risks and costs of cyber threats 193

The role of leadership in shaping
cybersecurity culture   194

Roles and responsibilities of different stakeholders 195

Defining stakeholder roles in cybersecurity   196

Interdepartmental collaboration
in cybersecurity   197

Engaging external stakeholders in
cybersecurity efforts   198

Promoting shared responsibility for cybersecurity 199

Creating a culture of cybersecurity awareness   200

Building cross-functional cybersecurity teams   201

Measuring and reinforcing cybersecurity
culture   202

Summary 203

12

Collaborating with Industry Partners and Sharing Threat Intelligence 205

The importance of collaboration and threat intelligence sharing 206

The imperative for collaborative defense   206

Mechanisms of threat intelligence sharing   207

Best practices in collaboration and sharing   208

Building trust and maintaining confidentiality in information sharing 209

Establishing trust among partners   210

Maintaining confidentiality
in information sharing   210

Balancing transparency and confidentiality   211

Leveraging shared threat intelligence for improved security 212

Integrating shared intelligence into security
operations   212

Collaborative incident response and recovery   213

Promoting shared responsibility for cybersecurity 215

Cultivating a culture of cybersecurity awareness   215

Engaging in public-private partnerships (PPPs)   216

Leveraging technology for collective defense   217

Summary 218

Index 219

Other Books You May Enjoy 230

Preface

Welcome to *Cybersecurity Strategies and Best Practices*, a guide for cybersecurity professionals to navigate the constantly evolving landscape of cybersecurity. With the advancement of technology, cyber adversaries are now using increasingly sophisticated tactics such as malware, ransomware, social engineering, and insider threats. This book will guide you through mitigating the risks associated with these evolving threats using case studies and industry best practices.

This book covers profiling adversaries, assessing weaknesses, and developing comprehensive strategies that align with business objectives. Organizations can mitigate risks and respond effectively to incidents by fostering security awareness and leveraging advanced technologies.

In today's interconnected world, cybersecurity is a necessity. Whether you're an experienced expert or new to the field, this book equips you with the necessary tools to protect data, systems, and reputation, ensuring a secure digital future. By the end of the book, you'll be well-equipped to safeguard your data, systems, and reputation, ensuring a secure digital future.

Who this book is for

This book is perfect for cybersecurity professionals with a foundational understanding of cybersecurity who seek to enhance their expertise in cybersecurity strategies and best practices by learning from real-world case studies that will help them align their organizational security measures with business objectives to combat the continuously evolving threat landscape.

What this book covers

Chapter 1, *Profiling Cyber Adversaries and Their Tactics*, provides an overview of different types of threat actors (e.g., nation-state and APT), their motivations (e.g., espionage, economic damage, or extortion), and the typical **tactics, techniques, and procedures** (**TTPs**) they employ.

Chapter 2, *Identifying and Assessing Organizational Weaknesses*, guides you through identifying and assessing vulnerabilities and weaknesses within your organization's enterprise network and cloud environment across endpoints, identities, networks, and cloud workloads.

Chapter 3, *Staying Ahead: Monitoring Emerging Threats and Trends*, focuses on the importance of staying up to date with emerging threats and trends in cybersecurity. The chapter will discuss the role of innovation and collaboration in staying ahead of the evolving threat landscape.

Chapter 4, Assessing Your Organization's Security Posture, teaches you how to evaluate your organization's overall security posture by considering technology, processes, and people. The chapter will discuss metrics to measure the effectiveness of security controls and the importance of maintaining a comprehensive and up-to-date inventory of assets.

Chapter 5, Developing a Comprehensive Modern Cybersecurity Strategy, focuses on creating a modern cybersecurity strategy that aligns with organizational objectives, considers current and emerging threats, and is adaptable to change. You will learn about key elements of a successful cybersecurity strategy, including risk management, digital forensics incident response, and security awareness programs.

Chapter 6, Aligning Security Measures with Business Objectives, explains the importance of aligning security measures with business objectives to ensure that cybersecurity initiatives support organizational goals. The chapter will discuss strategies for communicating the value of security investments to non-technical stakeholders and approaches for prioritizing security initiatives based on business impact.

Chapter 7, Demystifying Technology and Vendor Claims, aims to equip you with the knowledge and skills needed to critically evaluate technology and vendor claims. You will learn how to ask the right questions and strategies for making informed decisions when selecting cybersecurity products and/or services.

Chapter 8, Leveraging Existing Tools for Enhanced Security, focuses on helping you identify and optimize tools within your organization to enhance cybersecurity. You will learn about common tools and technologies that can be repurposed or integrated with other solutions to improve security posture.

Chapter 9, Selecting and Implementing the Right Cybersecurity Solutions, teaches you about selecting and implementing the proper cybersecurity solutions for your organization. The chapter will cover key factors to consider during the selection process, such as training, procedures, compatibility, scalability, usability, and best practices for successful implementation and integration.

Chapter 10, Bridging the Gap between Technical and Non-Technical Stakeholders, addresses the importance of effective communication and collaboration between technical and non-technical stakeholders in an organization. You will learn strategies for translating technical concepts into business language, fostering a security-aware culture, and building trust between different teams and departments.

Chapter 11, Building a Cybersecurity-Aware Organizational Culture, discusses the importance of developing a cybersecurity-aware organizational culture and provides strategies for building and maintaining such a culture. You will learn about the roles and responsibilities of different stakeholders, and how to promote a culture of shared responsibility for cybersecurity.

Chapter 12, Collaborating with Industry Partners and Sharing Threat Intelligence, discusses the importance of collaboration and sharing threat intelligence to improve the cybersecurity posture. You will learn about various threat intelligence-sharing platforms, frameworks, and best collaboration and information-sharing practices.

To get the most out of this book

You should have a foundational understanding of security concepts and tooling. However, before reading the book, no advanced knowledge of cybersecurity strategies or best practices is necessary.

Conventions used

There are a number of text conventions used throughout this book.

`Code in text`: Indicates code words in text, database table names, folder names, filenames, file extensions, pathnames, dummy URLs, user input, and Twitter handles. Here is an example: "Mount the downloaded `WebStorm-10*.dmg` disk image file as another disk in your system."

Bold: Indicates a new term, an important word, or words that you see onscreen. For instance, words in menus or dialog boxes appear in **bold**. Here is an example: "Select **System info** from the **Administration** panel."

> **Tips or important notes**
> Appear like this.

Get in touch

Feedback from our readers is always welcome.

General feedback: If you have questions about any aspect of this book, email us at `customercare@ packtpub.com` and mention the book title in the subject of your message.

Errata: Although we have taken every care to ensure the accuracy of our content, mistakes do happen. If you have found a mistake in this book, we would be grateful if you would report this to us. Please visit `www.packtpub.com/support/errata` and fill in the form.

Piracy: If you come across any illegal copies of our works in any form on the internet, we would be grateful if you would provide us with the location address or website name. Please contact us at `copyright@packt.com` with a link to the material.

If you are interested in becoming an author: If there is a topic that you have expertise in and you are interested in either writing or contributing to a book, please visit `authors.packtpub.com`.

Share Your Thoughts

Once you've read *Cybersecurity Strategies and Best Practices*, we'd love to hear your thoughts! Scan the QR code below to go straight to the Amazon review page for this book and share your feedback.

https://packt.link/r/1803230053

Your review is important to us and the tech community and will help us make sure we're delivering excellent quality content.

Download a free PDF copy of this book

Thanks for purchasing this book!

Do you like to read on the go but are unable to carry your print books everywhere?

Is your eBook purchase not compatible with the device of your choice?

Don't worry, now with every Packt book you get a DRM-free PDF version of that book at no cost.

Read anywhere, any place, on any device. Search, copy, and paste code from your favorite technical books directly into your application.

The perks don't stop there, you can get exclusive access to discounts, newsletters, and great free content in your inbox daily

Follow these simple steps to get the benefits:

1. Scan the QR code or visit the link below

https://packt.link/free-ebook/9781803230054

2. Submit your proof of purchase
3. That's it! We'll send your free PDF and other benefits to your email directly

1

Profiling Cyber Adversaries and Their Tactics

Cyber threats have become a critical component of our digital world. From state-sponsored hackers to rogue individuals, corporate spies, and organized crime units, these threat actors come in many forms and possess the skills and capacities to wreak havoc on our online infrastructure. Motivated by various objectives, such as financial gain or espionage, threat actors employ a complex array of **Tactics, Techniques, and Procedures** (**TTPs**) for their attacks. These tactics may include anything from phishing campaigns, malicious software, social engineering, and network intrusions to data manipulation or theft.

In this chapter, we will discuss the motivations and objectives of threat actors and explore some real-world examples of cyber-attacks. We will also look at the different types of TTPs used by threat actors and evaluate measures that can be taken to protect against them. Ultimately, the goal is for you to gain a better understanding of cyber threats and the actions necessary to secure your systems against malicious actors.

We will cover the following topics:

- Types of threat actors
- Motivations and objectives of threat actors
- Tactics, Techniques, and Procedures (TTPs)
- Real-world examples of cyberattacks and consequences

Types of threat actors

It is time for the next change in how security professionals approach not only building defenses but best practices for identifying, responding to, sustaining, and recovering from attacks. While, historically, it was all about building preventative defenses and even often assuming that the organization would never be targeted, at one point, it was understood that organizations must continuously assume breaches.

By assuming breaches, organizations prepare for the worst-case scenario because it is acknowledged that it's no longer *if* but *when* they will be targeted. However, now we must go to the next step, assume an attacker's mindset, and anticipate their next move while becoming more resilient. As seen in the following figure, as an industry, it's time to push into the stage of anticipation.

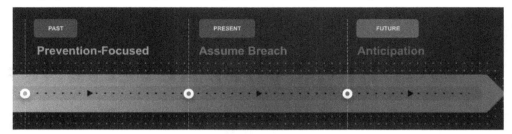

Figure 1.1 – Evolution of cybersecurity mindset

Putting oneself in the attacker's mindset is crucial as it allows for building effective incident response procedures, which can help ensure that all necessary steps are taken immediately following an attack. Furthermore, proactive measures such as implementing robust security controls, conducting continuous threat and vulnerability assessments, training security and end users on cyber hygiene best practices, and regularly testing your incident response plans are essential to any organization's modern cybersecurity strategy. However, all this can only be possible If we deeply understand the threat actors and the landscape. Remember that threat actors are also putting themselves into a defender's mindset and consider how incident response procedures might be modeled, the tooling you might have and how it's configured, and respective **Service-Level Agreements** (**SLAs**) with stakeholders. As defenders, we must understand who we are against; this will help us prioritize our defense strategy. Just as a chess player must study their opponent's moves to plan a winning strategy, defenders must understand their adversaries to prioritize their defense tactics effectively.

> *"If you know the enemy and know yourself, you need not fear the result of a hundred battles. If you know yourself but not the enemy, for every victory gained, you will also suffer a defeat. If you know neither the enemy nor yourself, you will succumb in every battle." — Sun Tzu, The Art of War*

Cybersecurity threats can be posed by a group or individual, including nation-state threat actors, hacktivists, cybercriminals, script kiddies, and **Advanced Persistent Threat** (**APT**) actors. Each type of threat actor has motivations and objectives in launching attacks against organizations or individuals. First, security professionals must understand the different kinds of threat actor categories so that, in the next step, it's possible to understand the TTPs used to attack systems and the potential consequences should the threat actor be successful. When diving into the different kinds of threat actors, it is critical to understand their unique motives, their available resources, and the methodology they use for their cyber attacks. Not all attacks are equal, and not all security controls will be adequate for all attacks.

Summary of threat actor categories

Threat Actor Categories	Description
Cybercriminals	Cybercriminals engage in cybercrime for financial gain while commonly using ransomware, identity theft, or phishing tactics.
Nation-State Actors	Nation-state actors are state-sponsored groups or individuals that often operate in the shadows and act on behalf of a government, typically seeking to disrupt critical infrastructure, steal sensitive data, or conduct cyber espionage.
Hacktivists or Activists	Hacktivists are cyber activists who use hacking techniques to promote a social or political cause, often targeting organizations or governments they perceive as unjust.
Insider Threats	Insider threats can be current or former employees, contractors, or partners with access to an organization's systems that misuse that access to cause harm, steal information, or commit other malicious acts.
Advanced Persistent Threats (APTs)	**Advanced Persistent Threats**, or **APTs** for short, are highly skilled and often well-funded threat actors, who engage in persistent and highly targeted attacks, using sophisticated tactics and techniques to infiltrate and maintain access to systems.
Cyber-Terrorists	Cyber-terrorists can be individuals or groups who use cyber attacks to cause chaos, fear, or physical harm, often driven by religious, political, or ideological motivations.
Script Kiddies	Script kiddies are inexperienced hackers who use readily available scripts or tools to carry out cyber attacks, often without a specific target or objective, seeking to cause disruption.
Private-Sector Offensive Actors (PSOAs)	**Private-Sector Offensive Actors** (**PSOAs**), also commonly referred to as hackers-for-hire, are highly skilled criminals who offer their services to the highest bidder, whether a corporation, criminal, or government organization.

Each type of threat actor presents unique challenges and requires tailored defense, identification, response, and recovery strategies to mitigate the risks they pose effectively. Therefore, let's dive deeper into each of these threat actor types.

Cybercriminals

Cybercriminals continuously hunt for vulnerabilities that can be exploited to gain unauthorized access to sensitive data, often for financial benefits. Cybercriminals manifest in different types, from individual cybercriminals who typically aim to compromise individual accounts to crime syndicates with an extensive global reach. Cybercriminals commonly leverage phishing campaigns, identity theft, and ransomware attacks to steal valuable information or extort money from their victims. As cybercriminals evolve and adapt, cybercriminals employ increasingly sophisticated techniques and tools to compromise security controls and penetrate personal and organizational systems. Some cybercriminals target vulnerable small businesses, while others seek to infiltrate large corporations, government agencies, or critical infrastructure.

Nation-state actors

Nation-state threat actors are among the most feared threat actor categories. The reason is that nation-state actors, the majority of the time, have significant resources to plan and execute large-scale and highly sophisticated cyber attacks. The majority of the threat actor groups that are sponsored by governments or state-affiliated entities operate in the shadows in complete secrecy and aim to steal sensitive information, disrupt critical infrastructure, or conduct cyber espionage operations. It's crucial to not take nation-state actors lightly due to their access to significant resources and vast networks, allowing them to launch massive global campaigns against any target quickly.

Hacktivists

Hacktivists, also called cyber activists, are not a new threat actor category but have existed since 1996. In 1996, **Cult of the Dead Cow** (**CDC**) members coined the term hacktivism. The CDC was an early hacktivist collective that exposed government secrets and fought for freedom of speech on the internet. Hacktivists use hacking techniques to promote social or political causes. Their victims are organizations and governments that hacktivists deem unjust, to bring attention to their cause.

A hacktivist group example is Anonymous, which has performed several cyberattacks over the years, including an operation against Scientology in 2008 and Operation Payback, which occurred after organizations including PayPal, MasterCard, and Visa imposed blockades on Wikileaks in 2010. In recent years, hacktivist groups have also increasingly performed **Distributed Denial-of-Service** (**DDoS**) attacks against websites belonging to the United States of America's law enforcement agencies. A recent example is during the massive protests following George Floyd's death. The motivations behind hacktivism vary vastly, depending on which hacktivist group is involved and what cause they stand for. While many hackers are financially motivated, others may be motivated purely by ideological reasons or a desire for justice and revenge.

Insider threats

Insider threats encompass various actors, including current and former employees, partners, contractors, and other trusted insiders with authorized access to corporate systems and sensitive data. Insider threats pose a significant challenge for security professionals, given that insider threats originate from individuals with authorized environmental access. Insider threats deliberately abuse their access privileges to cause damage or steal valuable and often highly classified information.

Furthermore, insider threats are especially dangerous given that these individuals often have intimate knowledge of the organization's policies, procedures, and infrastructure, which they can exploit to bypass security controls and avoid detection. Insider threat tactics are diverse and often insidious, with some actors leveraging social engineering techniques to manipulate their colleagues into granting them further access or revealing sensitive information. This may involve tactics such as impersonating an authority figure or exploiting the trust of fellow employees. By operating covertly, these individuals can carry out their malicious activities undetected for extended periods, potentially causing extensive harm to the organization.

Advanced Persistent Threats (APTs)

Advanced Persistent Threats (APTs) are a threat actor category that can conduct highly sophisticated and systematic cyber-attacks on specific targets, often to steal sensitive information or disrupt operations. APTs often leverage sophisticated techniques, including access to critical software vulnerabilities, unknown to the world, aka zero-days, to infiltrate and maintain access, through persistence, while often evading typical defenses and threat detection tools and processes. Once the APT is inside the target environment, they leverage various tactics, such as creating backdoors, installing keyloggers, and exfiltrating data, often in small batches but frequently to avoid detection. APTs are often well-resourced and highly motivated groups that can remain undetected for months or even years. Given APTs' ability to operate in stealth and establish persistence, it makes them, along with nation-state actors, among the most dangerous cyber threats today, which requires security professionals always to be vigilant and have robust security procedures and tooling in place to defend against them.

While gaining insights about the various APT groups is challenging, security professionals can leverage publicly available libraries such as the one from the MITRE organization, a not-for-profit that provides technical guidance for government agencies and developed the MITRE ATT&CK framework—a comprehensive knowledge base modeling cyber adversary tactics and techniques for threat detection and prevention. For detailed APT group TTPs, visit `https://attack.mitre.org/groups/`.

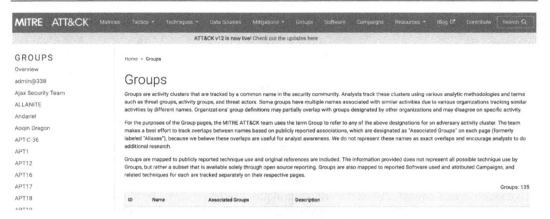

Figure 1.2 – MITRE ATT&CK Framework Groups

Some notable APT groups include the following:

Suspected Origin	Group Name	Insights
Russia	APT28	APT28 has been attributed to Russia's General Staff Main Intelligence Directorate (GRU) 85th Main Special Service Center (GTsSS) military unit 26165. In 2016, APT28 reportedly compromised the Hilary Clinton U.S. presidential campaign, the Democratic Congressional Campaign Committee, and the Democratic National Committee to interfere with the U.S. presidential election.
Russia	APT29	APT29 has been associated with Russia's Foreign Intelligence Service (SVR). It is believed that APT29 has been operating since at least 2008, primarily targeting government networks in Europe and NATO member countries, think tanks, and research institutes. In 2021, the US and UK governments attributed the SolarWinds supply chain attack to APT29.
Russia	FIN7	FIN7 is a Russian-based APT group that is financially motivated and has been active since 2013. FIN7 primarily targets hospitality sectors, retail, and restaurants and often deploys malware on point-of-sale (PoS) devices. Since 2020, FIN7 is suspected of having also started running their **Ransomware as a Service (RaaS)** underground.

Suspected Origin	Group Name	Insights
Russia	Sandworm	Sandworm has been associated with Russia's General Staff Main Intelligence Directorate (GRU) Main Center for Special Technologies (GTsST) military unit 74455. This specific APT group has been linked to several highly impactful cyber attacks, including in 2015 and 2016 against Ukrainian electric companies and governments, the 2019 worldwide NotPetya campaign, and the 2018 operation against the Organization for the Prohibition of Chemical Weapons.
China	APT1	APT1 is allegedly connected to the 2nd Bureau of the **People's Liberation Army (PLA) General Staff Department's (GSD's)** 3rd Department, known by its **Military Unit Cover Designator (MUCD)** as Unit 61398.
China	APT3	APT3 has been associated with China's Ministry of State Security. This APT group is responsible for high-profile campaigns such as Operation Clandestine Wolf or Operation Double Tap. Since 2015, APT3 commonly targets political organizations in Hong Kong.
China	APT17	APT17 is a China-based APT group that commonly targets U.S. government entities, the defense industry, law firms, mining companies, and non-government organizations and conducts network intrusion attacks.
Iran	APT33	APT33 is allegedly affiliated with the Iranian government. This APT group has been operating since 2013. APT33 has targeted organizations across multiple industries, however, often focusing on organizations in the United States, South Korea, and the Kingdom of Saudi Arabia.
Iran	APT34, OilRig	APT34, also commonly referred to as OilRig, is suspected to be linked to the Iranian Ministry of Intelligence and Security, which has been targeting Middle Eastern and international victims since 2014. This group typically carries out supply chain-based attacks, leveraging the trust relationship between organizations to attack their primary targets.
Iran	APT39	APT39 is using a front company named Rana Intelligence Computing; however, APT39 is attributed to the Iranian **Ministry of Intelligence and Security (MOIS)** and is believed to be chartered for cyber espionage on behalf of the Iranian government. This APT group primarily targets hospitality, academic, telecommunication, and travel industries across Europe, Africa, North America, and Asia.

Suspected Origin	Group Name	Insights
North Korea	APT37	APT37 is a state-sponsored cyber espionage APT group from North Korea. APT37 has been active since 2012 and targets victims primarily in South Korea but occasionally in Vietnam, Russia, Japan, Nepal, India, China, Kuwait, and Romania. APT37 is responsible for several global campaigns, including Operation Erebus, Evil New Year, and North Korean Human Rights.
North Korea	APT38	APT38 is attributed to the North Korean government and is one of their nation-state-funded threat actor groups, which has operated since at least 2014. APT38 is specialized in cyber financial operations and targets cryptocurrency exchanges, banks, casinos, and financial institutions.
Vietnam	APT32	Although not confirmed, security researchers believe APT32 is a Vietnam-based APT group that has operated since 2014. APT32 targets various industries and organizations, including foreign governments and journalists.

It is essential to know that APT groups employ sophisticated TTPs to infiltrate and maintain access to their targets' networks, often going undetected for extended periods. Defending against APT groups requires robust security controls, including continuous monitoring, threat intelligence, and incident response capabilities. Furthermore, it is critical to understand which types of APTs you will want to monitor more closely to ensure alignment in your defenses against them. It's essential to ask yourself, *"Which APTs are most likely to target my organization?"* and *"Which APTs most frequently target my industry and/or segment?"*.

Cyber-terrorists

Cyber-terrorists' primary charter is to perform cyberattacks that cause chaos, fear, and physical harm for religious, ideological, or political reasons.

Typically, cyber-terrorists target infrastructure systems, including transportation networks or power grids, which allows them to disrupt at a large scale. Occasionally, cyber-terrorists have been confirmed as launching cyberattacks against government entities or private organizations to gain access to sensitive data or extortion. It's essential to be aware that cyber-terrorists usually perform social engineering tactics such as spearphishing emails explicitly designed for high-value targets, from which they later deploy malicious code through web servers or other means of communication.

Script kiddies

Script kiddies are a type of threat actor that lacks technical knowledge and experience. Script kiddies typically use pre-made shelve exploits, malicious tools, and scripts to carry out their attacks rather than developing scripts or toolsets.

Furthermore, script kiddies often have no clear objective or target in mind but, instead, just continuously probe till they find a victim. They may carry out attacks simply for the sake of causing disruption or seeking attention from the media or online communities. However, despite their lack of expertise, it's crucial never to underestimate these attackers as they only need to succeed once, and security professionals must defend against them every single time. Therefore, script kiddies can still threaten organizations and individuals, as their attacks can cause disruption and compromise sensitive data.

Private-Sector Offensive Actors (PSOAs)

Private-Sector Offensive Actors (**PSOAs**) offer their hacking skills to their highest bidder, which is why they are also commonly called hackers-for-hire. PSOAs offer their services to government agencies, corporations, and criminal organizations. These hackers typically employ advanced TTPs, such as social engineering and malicious software, to infiltrate organizations' networks and extract valuable data. PSOAs can cause significant economic damage if left unchecked; hence, organizations must understand this threat actor's motivations and TTPs to defend against them properly.

As the next step, it's critical to understand the motivations and objectives of the different threat actor categories.

Motivations and objectives of threat actors

Cybercriminals are often motivated by financial gain and use methods including identity theft, phishing, or ransomware. Nation-state actors, on the other hand, pursue geopolitical ambitions through cyber espionage. Cyber-terrorists and hacktivists leverage their criminal skills to promote a specific agenda or inflict damage upon perceived enemies.

As security professionals, it is critical to develop targeted countermeasures that protect valuable assets from potential threats. It always goes back to knowing your enemy to build the best defense strategy. With that, security professionals can assess risk and take the appropriate measures to maintain a secure digital environment. To further illustrate the various motivations and objectives, here's a brief list of examples for each of the threat actor categories:

- **Cybercriminals:** The Carbanak group is a cybercriminal syndicate that has operated since at least 2013. Some security researchers believe that Carbanak might be associated with Cobalt Group and FIN7, as all use the Carbanak malware. The Carbanak group is responsible for stealing over $1 billion from financial institutions worldwide, using tactics such as spear-phishing emails, malware, and data exfiltration.

  Source: `https://www.europol.europa.eu/newsroom/news/mastermind-behind-eur-1-billion-cyber-bank-robbery-arrested-in-spain`

- **Nation-state actors:** APT29, a threat actor group associated with Russia's Foreign Intelligence Service (SVR), has targeted various government and private sector organizations, including the **Democratic National Committee** (**DNC**), during the 2016 U.S. presidential election.

Source: https://www.gov.uk/government/news/russia-uk-and-us-expose-global-campaigns-of-malign-activity-by-russian-intelligence-services

- **Hacktivists:** Anonymous is an organized collective of cyber activists that target organizations and governments they perceive as unjust. Anonymous has launched cyberattacks against various government and corporate entities, including the Church of Scientology, PayPal, and the FBI.

- **Insider threats:** The most prominent case of insider threats to this day is Edward Snowden, a former **National Security Agency (NSA)** contractor who leaked classified information about global surveillance programs, exposing widespread government surveillance practices.

- **Advanced Persistent Threats (APTs):** APT36, a Pakistan-aligned threat actor, expands interest in the Indian Education Sector. Security researchers at SentinelLabs discovered that APT36 has been introducing OLE embedding into its typically used techniques for staging malware, from lure documents and version changes to the implementation of Crimson RAT, indicating the ongoing evolution of APT36's tactics and malware arsenal.

Source: https://www.sentinelone.com/labs/transparent-tribe-apt36-pakistan-aligned-threat-actor-expands-interest-in-indian-education-sector/

- **Cyber-terrorists:** The Cyber Caliphate, a group claiming affiliation with ISIS, has conducted cyberattacks against media organizations, government websites, and social media accounts, spreading propaganda and promoting terrorist activities.

- **Script kiddies:** In 2000, a 15-year-old Canadian hacker known as "Mafiaboy" launched a series of high-profile DDoS attacks against major websites such as Yahoo!, eBay, and Amazon, causing widespread disruption and significant financial losses.

Source: https://www.heraldsun.com.au/news/victoria/former-teen-hacker-mafiaboy-was-hunted-by-the-fbi-but-now-fights-cyber-crime/news-story/7f322b8403d023d5c7de662fd14072fb

- **Private-sector offensive actor (PSOA):** HackingTeam was an Italian company that provided offensive intrusion and surveillance software to governments, law enforcement agencies, and corporations. They have been criticized for selling their services to authoritarian regimes, enabling surveillance and monitoring of dissidents and political opponents.

Tactics, Techniques, and Procedures (TTPs)

Tactics, Techniques, and Procedures (TTPs) represent the modus operandi of threat actors as they engage in cyberattacks or other malicious activities. These TTPs provide a structured framework for understanding and categorizing the behavior and methods employed by threat actors, enabling cybersecurity professionals to identify patterns, anticipate threats, and develop adequate security controls and, ultimately, a robust information security program. Security professionals need to comprehend the TTPs employed by threat actors as it will help to identify patterns, anticipate threats, and develop tailored countermeasures. Tactics refer to an attack's overarching goals or objectives, such as data exfiltration or gaining unauthorized access. Threat actors utilize various techniques to achieve their tactics, including exploiting software vulnerabilities, launching social engineering attacks such as phishing, and executing malware attacks.

The MITRE ATT&CK Framework is a comprehensive knowledge base and model for threat actor behavior and their TTPs. This framework is one of the most widely used and comprehensive TTP frameworks in the cybersecurity industry, and today, many security teams use it as their primary resource for TTPs. Given the significance of the MITRE ATT&CK Framework, it is critical to understand it better. The framework is developed by MITRE, which is a not-for-profit organization. The ATT&CK Framework (for Adversarial Tactics, Techniques, and Common Knowledge) provides an in-depth understanding of the TTPs employed by threat actors across various stages of a cyberattack. By compiling this information in a structured manner, the framework enables security professionals to analyze, anticipate, and defend against a wide range of cyber threats.

The ATT&CK Framework is organized into a series of matrices that cover different aspects of the cyber threat landscape, such as Enterprise, ICS (Industrial Control Systems), and Mobile. Each matrix is divided into tactics, which represent the high-level goals of an attacker, and techniques, which are the methods or approaches used to achieve these goals. Within each technique, there are sub-techniques that provide further granularity and detail about specific adversary actions. You can find the MITRE ATT&CK Matrix for Enterprise at https://attack.mitre.org/matrices/enterprise/.

One key aspect of the MITRE ATT&CK Framework is its focus on the entire lifecycle of a cyberattack. The framework consists of several stages: initial access, execution, persistence, privilege escalation, defense evasion, credential access, discovery, lateral movement, collection, command and control, and exfiltration. Understanding the MITRE ATT&CK Framework can help security professionals build their threat intelligence program and improve their incident response, security posture, security operations, red teaming, and vulnerability management abilities.

Besides the MITRE ATT&CK Framework, MITRE also provides the MITRE ATT&CK Evaluations, a comprehensive assessment of security products and services against real-world cyber threats. These evaluations can help security professionals better understand which security products and services best suit their organizations' needs and avoid falling for false vendor marketing claims. This independent evaluation measures the performance of cybersecurity products and services in detecting, preventing, and responding to sophisticated attack scenarios that are modeled after TTPs used by advanced threat actors. The following figure is from the MITRE Engenuity ATT&CK Evaluation website showing recent enterprise evaluations: https://attackevals.mitre-engenuity.org/enterprise/.

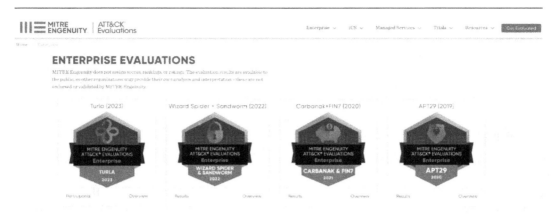

Figure 1.3 – MITRE ATT&CK Enterprise Evaluations

The evaluation results offer a standardized benchmark for organizations to compare and contrast various security solutions, empowering them to make informed decisions when selecting tools and technologies to enhance their cybersecurity posture. The following figure is from the MITRE ATT&CK Evaluation website, showing how they monitor various participating vendors – in this example, SentinelOne.

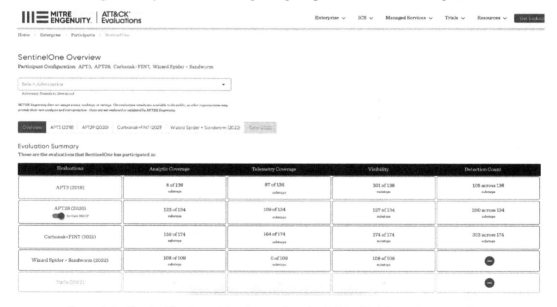

Figure 1.4 – SentinelOne's participation card on the MITRE ATT&CK Evaluation website

Furthermore, the ATT&CK Evaluation fosters an environment of continuous improvement among security vendors as they strive to refine their offerings based on the insights gained from the evaluation process. By leveraging the MITRE ATT&CK Evaluation, organizations can better understand the strengths and weaknesses of their security infrastructure, ultimately leading to more effective strategies and a heightened level of protection against evolving cyber threats.

In addition to understanding the different threat actor categories and their motivations, objectives, and commonly used TTPs, it is essential to dive deeper into some real-world cyber-attack examples as it will provide security professionals with needed guidance when considering one's defenses.

Real-world examples of cyberattacks and consequences

This section discusses real-world examples of cyberattacks attributed to nation-state actors, APT groups, and other malicious entities such as cybercriminals.

Nation-state actors: NotPetya attack (2017)

Sandworm, which is an APT group attributed to Russia's General Staff Main Intelligence Directorate (GRU) Main Center for Special Technologies (GTsST) military unit 74455, executed a highly destructive cyberattack known as NotPetya, which targeted various organizations across the globe and resulted in damages that exceeded $10 billion. Although NotPetya initially appeared as a ransomware campaign, it was later discovered that the primary goal was not financial gain but widespread disruption and destruction. Security researchers confirmed that NotPetya was wiper malware designed to irreversibly corrupt and overwrite data on infected systems. It is believed that NotPetya was executed by Sandworm due to the ongoing conflict between Russia and Ukraine. The following diagram provides insights into NotPetya's attack kill chain.

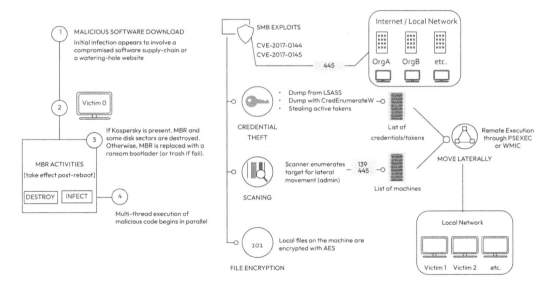

Figure 1.5 – NotPetya Attack Kill Chain

It is possible to identify several MITRE ATT&CK TTPs as part of the NotPetya cyberattack:

- T1190: Exploit Public-Facing Application (EternalBlue and EternalRomance were used to exploit SMB vulnerabilities in Windows)

- T1210: Exploitation of Remote Services (NotPetya leveraged SMB vulnerabilities to spread laterally across networks)

- T1003: OS Credential Dumping (Mimikatz was used to extract credentials from the memory of infected systems)

- T1078: Valid Accounts (Stolen credentials were used for lateral movement and further system access)

- T1485: Data Destruction (NotPetya overwrote and corrupted files, resulting in irreparable data loss)

- T1490: Inhibit System Recovery (The malware modified the MBR, preventing systems from booting up and complicating recovery efforts)

- T1193: Spearphishing Attachment (NotPetya was initially delivered via a malicious email attachment in a software update for Ukrainian tax software)

Source: `https://www.wired.com/story/notpetya-cyberattack-ukraine-russia-code-crashed-the-world/`

Nation-state actors: SolarWinds supply chain attack

Discovered in December 2020, the SolarWinds supply chain attack was a highly sophisticated and far-reaching cyber espionage campaign that targeted government agencies and private companies worldwide. The SolarWinds attack is believed to be linked to APT29, a nation-state threat actor related to the Russian Foreign Intelligence Service (SVR). This attack compromised the software development infrastructure of SolarWinds, from which the threat actor could insert a backdoor, known as "SUNBURST," into a legitimate update of the SolarWinds Orion platform that was then distributed to thousands of SolarWinds customers. With this approach, the threat actor could gain remote access to the networks of organizations that installed the compromised update. Its victims included the U.S. Departments of Treasury, Commerce, and Homeland Security and private sector companies such as Microsoft. The following diagram is a visual aid to better understand the kill chain.

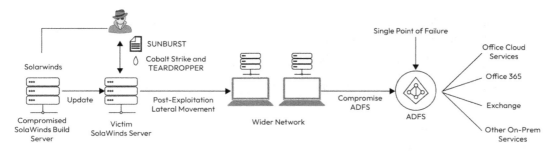

Figure 1.6 – SolarWinds attack kill chain

The SolarWinds supply chain attack involved several TTPs that can be mapped to the MITRE ATT&CK Framework. Some of the critical TTPs associated with the attack include the following:

- T1195.002: Supply Chain Compromise: Compromise Software Supply Chain (The attackers compromised the SolarWinds Orion platform update by inserting the SUNBURST backdoor)

- T1059: Command and Scripting Interpreter (The SUNBURST backdoor executed commands on the compromised systems)

- T1547: Boot or Logon Autostart Execution (The SUNBURST backdoor established persistence by creating a scheduled task that started when the system booted or the user logged in)

- T1134: Access Token Manipulation (The attackers leveraged the permissions of the SolarWinds Orion platform to escalate their privileges on the compromised systems)

- T1070: Indicator Removal on Host (The attackers removed forensic artifacts and logs to hide their activities)

- T1027: Obfuscated Files or Information (The SUNBURST backdoor employed various obfuscation techniques to evade detection)

- T1003: OS Credential Dumping (The attackers accessed credentials from the compromised systems to facilitate lateral movement)

- T1087: Account Discovery (The attackers gathered information about user accounts on the compromised systems)

- T1016: System Network Configuration Discovery (The attackers obtained network configuration information to understand the target environment better)

- T1021: Remote Services (The attackers used remote services such as SMB and RDP to move laterally across the target network)

- T1114: Email Collection (The attackers accessed the email accounts of targeted organizations to collect sensitive information)

- T1005: Data from Local System (The attackers collected data from the compromised systems)

- T1041: Exfiltration Over C2 Channel (The collected data was exfiltrated to the attackers' command and control servers)

Source: `https://www.sentinelone.com/labs/solarwinds-sunburst-backdoor-inside-the-apt-campaign/`

Cybercriminals: WannaCry ransomware attack (2017)

The WannaCry attack has been attributed to Lazarus Group which is a nation-state threat actor group from North Korea's Reconnaissance General Bureau department. WannaCry seems to have two primary objectives: to cause chaos and, secondly, to achieve financial gain. WannaCry was a ransomware attack in May 2017 and a large-scale global campaign that impacted over 200,000 endpoints across 150 countries. WannaCry targeted Microsoft Windows-based endpoints and encrypted users' files, demanding a ransom payment in Bitcoin to exchange the decryption key. The rapid and widespread nature of WannaCry led to significant disruptions in various sectors, including healthcare, finance, and logistics, causing an estimated economic loss of several billion dollars.

WannaCry leveraged the EternalBlue exploit, which targeted a vulnerability in the Windows **Server Message Block (SMB)** protocol. Like NotPetya, the NSA initially discovered the EternalBlue exploit and later leaked it by the Shadow Brokers. The exploit enabled the ransomware to spread rapidly across networks, infecting vulnerable systems without user interaction. Microsoft released a patch for the vulnerability (MS17-010) in March 2017, and many organizations had not yet applied the update, leaving their systems exposed to the attack. The following diagram visually helps to better understand the kill chain.

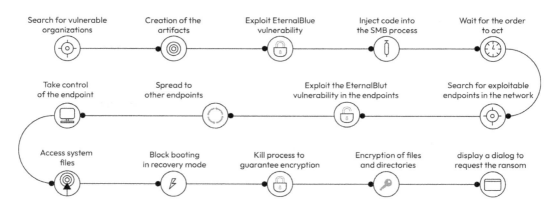

Figure 1.7 – WannaCry attack kill chain

Here's how the WannaCry attack can be mapped to MITRE ATT&CK TTPs:

- T1190: Exploit Public-Facing Application (EternalBlue was used to exploit the SMB vulnerability in Windows)

- T1486: Data Encrypted for Impact (WannaCry encrypted users' files, demanding a ransom payment for decryption)

- T1210: Exploitation of Remote Services (WannaCry leveraged the EternalBlue exploit for rapid propagation and lateral movement)

- T1490: Inhibit System Recovery (Victims were instructed to pay the ransom in Bitcoin to receive the decryption key)

- T1192: Spearphishing Link (Initial infection vectors may have included malicious email links)

Source: `https://www.reuters.com/article/us-cyber-attack-idUSKCN18B0AC`

Cybercriminals: Colonial Pipeline ransomware attack

The Colonial Pipeline ransomware attack was a high-profile cyber incident that took place in May 2021. The attack targeted the Colonial Pipeline, the largest fuel pipeline in the United States. The ransomware attack, carried out by a cybercriminal group known as DarkSide, encrypted the company's IT systems and demanded a ransom payment in exchange for the decryption key.

In response to the attack, Colonial Pipeline shut down its pipeline operations to prevent the ransomware's spread and ensure its systems' safety. This decision led to widespread fuel shortages and a temporary spike in fuel prices throughout the United States. Eventually, the company paid the attackers a ransom of approximately $4.4 million in Bitcoin to regain access to its systems.

The Colonial Pipeline attack highlights the increasing threat of ransomware attacks against critical infrastructure and the potential cascading effects these attacks can have on society. The incident also underscores the importance of robust cybersecurity measures and incident response plans for organizations in every sector. The following diagram showcases the kill chain.

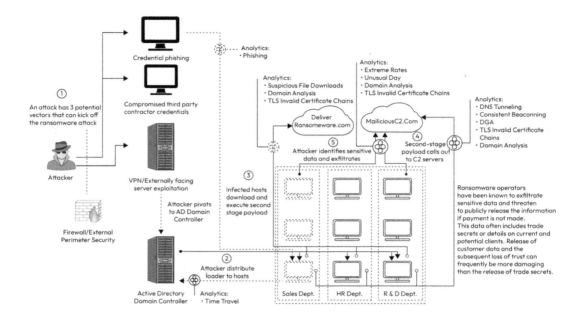

Figure 1.8 – Colonial Pipeline attack kill chain

Here's how the Colonial Pipeline attack can be mapped to MITRE ATT&CK TTPs:

- T1486: Data Encrypted for Impact (DarkSide ransomware encrypted the company's IT systems, demanding a ransom payment for decryption)

- T1490: Inhibit System Recovery (The company paid the ransom in Bitcoin to regain access to its systems)

- T1566: Phishing (DarkSide is known to use phishing emails as an initial infection vector, though the exact method used in the Colonial Pipeline attack is unclear)

- T1003: OS Credential Dumping (DarkSide ransomware often attempts to steal credentials from the compromised systems)

Source: https://www.bloomberg.com/news/articles/2021-05-13/colonial-pipeline-paid-hackers-nearly-5-million-in-ransom

Summary

In this chapter, you learned about the various threat actor types, uncovered some of the most damaging real-world cyber-attacks and how they were performed, and looked at the importance of the MITRE ATT&CK Framework, which provides a detailed profile of the typical cyber attack stages, and the applicable TTPs, with which organizations can learn better about the way threat actors operate, thus being better informed on how to bolster their cyber resilience.

In the next chapter, we will look at how we can identify and assess organizational weaknesses.

2
Identifying and Assessing Organizational Weaknesses

The cornerstones of any successful cybersecurity strategy are identifying and effectively assessing organizational weaknesses as well prioritizing business needs and roadmaps. With the rapidly evolving threat landscape and continuous increase of the attack surface and sheer volume of attacks itself, we must be able to make faster, smarter decisions. Weaknesses can span from unpatched software to negligent insider behavior, which can create exploitable gaps in security. Utilizing techniques such as compliance and regulatory requirements, business needs, emerging new technologies and threats, vulnerability assessments, penetration testing, and threat modeling help identify and assess these weaknesses. Additionally, cybersecurity strategies also should be defined in a way that meets future business growth and enhancement transitions.

Vulnerability scanning and penetration testing are critical components of a robust cybersecurity assessment framework. The former identifies potential points of exploitation in a system or network, while the latter simulates cyberattacks to test resilience. Risk assessments quantify or qualify the potential impacts of identified vulnerabilities. It's a crucial process that enables an organization to understand the consequences of exploited vulnerabilities and facilitates better decision-making around cybersecurity investments and strategies.

Post-assessment, it's crucial to prioritize and remediate weaknesses, which involves developing and executing a mitigation plan. Prioritization typically depends on factors including the criticality of the system, the potential impact of a breach, and the exploitability of the vulnerability.

By adhering to these practices, organizations can build a more resilient cyber defense system, ensuring business continuity and integrity of their information assets.

Understanding organizational weaknesses and vulnerabilities

Understanding the differences between organizational weaknesses and vulnerabilities is paramount to forming an effective cybersecurity strategy. Weaknesses are generally flaws or deficiencies in a system that can lead to its compromise, while vulnerabilities denote weaknesses in software that outside actors can exploit. Addressing these issues might require patching a piece of software and introducing better security policies, as well as user awareness and training initiatives.

While technical problems are a risk, process-related weaknesses such as inadequate security policies or incident response plans must also be considered. Moreover, human-based vulnerabilities such as employee unawareness can open an organization to social engineering attacks. Organizations must remain committed to understanding and defending against organizational weaknesses and vulnerabilities as the threat landscape changes. Doing so will enable them to build a comprehensive, robust cybersecurity strategy.

Types of organizational weaknesses

Let's explore the different types of organizational weaknesses. While there might be other ways to categorize them, when looking at organizational weaknesses from a 50,000-foot perspective, it boils down to three categories: technical, process, and human.

	Technical: Software, network, and hardware vulnerabilities can lead to technical weaknesses. Outdated hardware or software (e.g., firmware, operating systems, applications, etc.) that are not patched and secured or systems incorrectly configured can be a major security issue. For example, operating systems running older software versions without the most recent security updates can cause significant problems for computer users and networks. Ensuring all components are up to date with the latest security patches is essential for protecting against technical weaknesses. Additionally, all hardware installations should be securely implemented and network endpoints adequately protected to avoid potential vulnerabilities.
	Process: Organizations need adequate security policies and well-defined change management processes. Without which the organizations are left vulnerable to various threats. This could be anything from inadequate backup procedures to an insufficient incident response strategy in the event of a ransomware attack. While organizations must be prepared for such disasters, they need more than just a robust disaster recovery plan; they need to ensure they have the necessary protocols and procedures to respond quickly and effectively to potential incidents.

	Human: Humans are prone to mistakes, a fact that can lead to security incidents. This can be due to personnel lacking cybersecurity education, leaving them vulnerable to social engineering techniques such as phishing scams. It is also possible for insiders, whether by malicious intent or accident, to unwittingly cause significant security breaches. To prevent this, organizations must prioritize educating their staff on cybersecurity protocols and strategies and ensuring strict regulations are in place.

While these categories help structure our understanding of weaknesses, it's essential to remember that they often interact. For instance, a technical weakness can be exploited due to a process weakness (such as a lack of patch management) facilitated by a human weakness (perhaps clicking on a phishing link). This interconnectedness makes addressing all weaknesses vital to a comprehensive cybersecurity strategy.

Types of organizational vulnerabilities

Let's look closer into what types of organizational vulnerabilities exist. Similar to organizational weaknesses, there are many variations. We can categorize them into software, hardware, and network vulnerabilities. Let's explore these categories and consider practical examples to understand them better.

	Software vulnerabilities: This type of vulnerability allows malicious actors to break into a system and cause harm. To prevent such threats from occurring, it is critical to ensure that all applications are up-to-date with the latest security patches and fixes. As an example, in 2017, the WannaCry ransomware attack exploited a flaw in Microsoft's Server Message Block protocol that, if not patched, could have allowed an attacker to access the system. WannaCry is suspected to have spread to 150 countries, and the cybercrime caused an estimated $4 billion in losses across the globe.
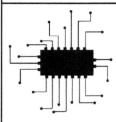	Hardware vulnerabilities: These are weaknesses in the physical components of a system that can lead to data leakage and theft. In 2018, two major hardware security flaws, Spectre and Meltdown, were discovered to affect modern AMD, Intel, and ARM processors. These vulnerabilities allowed malicious programs to access sensitive information stored in the computer's processor by exploiting its speculative execution feature. As a result, virtually all devices running on these processors were vulnerable to attacks.
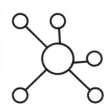	Network vulnerabilities: Vulnerabilities in network architecture and protocols can make systems susceptible to malicious attacks if configurations are left unsecured. For example, a Wi-Fi network that has not been adequately secured with encryption could easily be accessed by attackers, who can intercept traffic and steal confidential information.

As security professionals, it is crucial to be aware of the organization's environment's vulnerabilities. Knowing how these security flaws can be utilized maliciously is essential in implementing effective defensive techniques. Organizations should prioritize practices such as patching software regularly and ensuring secure configurations when it comes to network settings, as these measures can significantly reduce the chances of an attacker successfully exploiting a vulnerability.

Real-world examples

The global logistics company Maersk experienced a cyberattack in 2017 called NotPetya, triggered by a software vulnerability in their accounting software. This cyberattack resulted in the shutdown of 76 port terminals worldwide, taking Maersk two grueling weeks to restore its systems and costing an estimated $300 million.

Similarly, the 2017 Equifax breach compromised the sensitive data of approximately 147 million consumers when attackers exploited an unpatched Apache Struts web application vulnerability. This incident incurred major reputational damage and legal repercussions, with a whopping $575-million settlement.

The 2020 SolarWinds hack further highlighted the consequences of supply chain weaknesses, as hackers infiltrated SolarWinds' software development process and inserted a backdoor into an update for over 18,000 customers.

These examples demonstrate that managing organizational weaknesses and vulnerabilities is essential to mitigating damage and avoiding hefty costs. As such, it is crucial to maintain robust security protocols across all digital supply chain points and build an effective cybersecurity framework that promptly identifies, assesses, and remediates any vulnerabilities.

This is an essential lesson for all organizations to remember—the cost of not adequately addressing weaknesses and vulnerabilities can be immense. Organizations must prioritize the development of secure software solutions, protecting their digital supply chain, and mitigating human vulnerabilities to protect themselves from future cyberattacks.

Companies can proactively address security threats by adequately identifying and mitigating organizational weaknesses and vulnerabilities before they become damaging incidents.

Effective vulnerability management is essential for maintaining a strong cybersecurity posture. It enables businesses to identify risks associated with new technologies, keep ahead of emerging threats, and ensure business continuity in today's increasingly digital world. With proper implementation, organizations can rest assured that their critical assets are safe from malicious actors and prepared to address any security vulnerabilities quickly and efficiently.

Organizations face various vulnerabilities in their systems, which spans across software, hardware, and network vulnerabilities that can be exploited by threat actors. Instances such as the WannaCry ransomware attack, NotPetya, the Spectre and Meltdown hardware flaws, and insecure network configurations underscore the need for robust security measures. The high-profile attacks on Maersk, Equifax, and SolarWinds highlight the potential damage and financial costs of these vulnerabilities.

Therefore, it's crucial for organizations to proactively identify and mitigate these vulnerabilities, maintaining secure software solutions, protecting their digital supply chain, and training their staff to avoid cyberattacks. In doing so, companies can ensure their essential assets are protected and can deal with security threats swiftly and effectively.

Techniques for identifying and assessing weaknesses

Identifying and assessing systems and processes' weaknesses is integral to maintaining a secure environment. This helps detect possible points of exploitation and inform the development of effective security strategies.

Identification involves finding potential threats that could be exploited by malicious actors, such as outdated software, insecure configurations, insufficient policies, and even human factors such as a lack of awareness about cybersecurity. Assessment involves evaluating the identified risks to understand their impact and likelihood of exploitation, including severity ratings, the probability of exploitation, and the potential consequences.

Various techniques are available for these activities, from security audits and vulnerability assessments to penetration testing and social engineering tests. The method will depend on the organization's industry, the sensitivity of the data handled, the size of an organization, and the threat landscape.

By regularly identifying and assessing weaknesses within their systems and processes, organizations can effectively detect potential threats while minimizing their impacts if a successful attack occurs. This can help them remain one step ahead of cybercriminals and reduce the chances of a successful attack.

Security audits

Security audits should always be considered as they are essential for assessing and identifying flaws in an organization's IT protocols, systems, and policies. This is achieved by examining how well existing requirements and criteria are being met within the company.

Internal audits are conducted by a company's personnel or hired **subject matter experts** (**SMEs**) and focus on identifying weaknesses, such as outdated technology, misconfigurations, or non-conformity with internal rules. On the other hand, external audits are conducted by third-party organizations. Audits are often required to adhere to specific regulations such as ISO 27001, which deals with the overall management of information security, or the **Payment Card Industry Data Security Standard** (**PCI DSS**). It is important to be aware that government bodies can demand regulatory audits to ensure that regulations such as the **Health Insurance Portability and Accountability Act** (**HIPAA**) for healthcare firms or the **General Data Protection Regulation** (**GDPR**) for firms that manage **European Union** (**EU**) citizens' information are respected and that organizations are in compliance with them.

Furthermore, depending on the type of the business and its industry scope, additional regulatory compliance based on its geographic location may be applied as well. Hence, organizations define information security policies and standards accordingly to meet their own internal information security requirements as well as the regulatory requirements they are obliged to adhere to.

Security systems and processes require regular check-ups to identify weak points that could be exploited by threat actors. This process includes finding potential risks, such as insufficient security policies or outdated software, and assessing these risks based on their severity and likelihood of exploitation. Regular internal and external security audits are crucial to identify areas of improvement and ensure the organization complies with various regulations. These measures significantly reduce the risk of data breaches, keeping the organization one step ahead of potential threats.

Vulnerability assessments

Vulnerability assessments are critical to any organization's information security strategy, as they provide an in-depth analysis of weaknesses across their digital estate, including systems, networks, and infrastructure. These assessments can be conducted through automated scanning and manual reviews. Vulnerability management starts with asset discovery, where organizational assets are identified and cataloged. Next, vulnerability scanning is conducted to detect security weaknesses within the system. Following this, a vulnerability assessment is carried out, involving the evaluation and prioritization of the vulnerabilities based on their potential risk. The final step is vulnerability remediation, where solutions are applied to fix or mitigate the detected vulnerabilities, thereby enhancing the security posture of the organization.

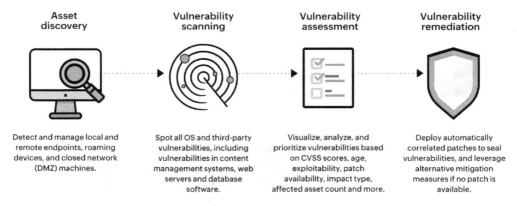

Figure 2.1 – Step-by-step vulnerability assessment process

Automated scanning involves running specialized tools, such as commercial software (e.g., Tenable Nessus, Qualys, or Rapid7 Nexpose) or open source products against databases of known vulnerabilities such as the **Common Vulnerabilities and Exposures** (CVE) list. These tools generate reports with details about the detected vulnerabilities and the recommended remediations.

Manual reviews involve security professionals thoroughly reviewing systems and processes to identify potential weaknesses that automated tools may miss. Due to the automation, additional vetting may be required to perform the next level of risk assessment and false-positive review to minimize the impact on operations. As part of this review, additional inputs from threat intelligence sources, targeted system threat landscapes, and system criticality could enhance the efficiency of the risk assessment process.

Once vulnerabilities are identified, they must be prioritized according to their severity, the sensitivity of the affected system, and the potential impact of a breach. This is an essential step, as it's important to acknowledge that there will always be vulnerabilities. At the same time, regardless of the organization's size, we always need to prioritize the workload. This prioritization helps organizations effectively allocate resources to address the most critical vulnerabilities first. By performing regular vulnerability assessments, organizations can keep their security posture up to date and minimize the risk of exploitation by attackers for malicious purposes.

Organizations should ensure their vulnerability assessment program is comprehensive enough to comply with applicable laws and regulations while providing sufficient protection against potential threats. This can involve leveraging specialized tools for automated scanning and engaging qualified personnel for manual reviews as part of a well-rounded approach to security evaluation. When done correctly, vulnerability assessments can go a long way in improving organizational cybersecurity.

By taking the necessary steps to assess and remediate vulnerabilities, organizations can significantly reduce their risk of being exploited by attackers, enhancing their security posture, and staying compliant with applicable regulations.

Vulnerability assessments help organizations identify and fix security weaknesses in their digital estate, which is critical for their cybersecurity strategy. This process involves identifying and cataloging all digital assets, scanning them for any potential vulnerabilities, evaluating these vulnerabilities, and then applying appropriate solutions to resolve them. Both automated tools and manual reviews by security professionals are used, and vulnerabilities are prioritized based on their severity and potential impact. Regular assessments enable organizations to stay updated on their security status and lower the risk of cyberattacks. Essentially, these assessments help organizations strengthen their digital defenses and stay in line with relevant laws and regulations.

Threat modeling

Threat modeling is a proactive approach to security that enables organizations to anticipate and prepare for potential cyberattacks. At its core, through threat identification, analysis, and risk assessment, organizations can determine which threats pose the most significant risks and develop strategies to mitigate them. This approach helps organizations to proactively anticipate and prepare for attacks rather than just reacting to security incidents.

One widely recognized methodology is **STRIDE** (which stands for **Spoofing, Tampering, Repudiation, Information Disclosure, Denial of Service, and Elevation of Privilege**), developed by Microsoft. This approach focuses on the types of attacks that could occur and helps organizations develop targeted defense strategies.

Threat	Desired Security Property
Spoofing	Authentication
Tampering	Integrity
Repudiation	Non-repudiation
Information disclosure	Confidentiality
Denial of service	Availability
Elevation of privilege	Authorization

Another model is **DREAD** (short for **Damage Potential, Reproducibility, Exploitability, Affected Users, and Discoverability**). This model quantifies each threat's risk level to prioritize mitigation efforts.

Threat	Desired Security Property
Damage	How bad would the attack be?
Reproducibility	How easy is it to reproduce attack?
Exploitability	How easy is it to recreate the attack?
Affected users	How many users could be impacted?
Discoverability	How easy is it to discover the attack?

The **Process for Attack Simulation and Threat Analysis (PASTA)** model is a more complete seven-step process combining threat identification and risk assessment.

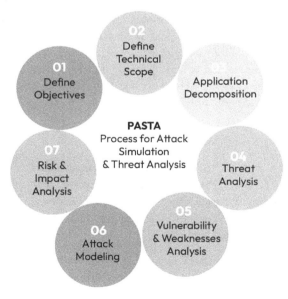

Figure 2.3 – The seven stages of the PASTA model

The best way for an organization to embrace threat modeling is by creating a proactive security culture. Teams should be encouraged to continuously monitor their systems and look for potential threats, such as new vulnerabilities or malicious actors. This will help organizations stay ahead of the ever-evolving digital threat landscape and better defend against cyberattacks.

Threat modeling helps organizations predict and prepare for potential cyber threats. It involves identifying potential threats, analyzing them, and assessing their risks to design defense strategies. Different models exist for this, such as STRIDE from Microsoft, which outlines types of attacks, DREAD, which scores the risk level of each threat, and PASTA, a comprehensive seven-step process that combines threat identification and risk assessment. To effectively use threat modeling, organizations need to foster a proactive security culture, encouraging teams to constantly monitor their systems for possible threats such as new vulnerabilities or malicious activity. This approach allows organizations to stay on top of the rapidly changing digital threat landscape and defend against cyberattacks more effectively.

Penetration testing

Penetration testing, more commonly known as 'pen testing,' is an authorized and proactive method of identifying security vulnerabilities in a system by simulating a cyberattack. Whereas vulnerability assessments are used to identify weaknesses, penetration tests go one step further by actively attempting to exploit these weaknesses to assess the potential damages should there be a breach.

Pen tests can come in many forms, including black-box testing, which mimics an external attacker without any prior knowledge of the system; white-box testing, which replicates an insider attack with a comprehensive understanding of the system; and grey-box testing, which is a combination of the two and provides a balanced approach to detecting potential vulnerabilities.

Once completed, a penetration test wraps up by creating a detailed report outlining all discovered vulnerabilities, the data accessed, and the recommended remediation actions. Tools that are highly popular when carrying out pen tests include Metasploit for developing and executing exploit code against target machines and Burp Suite for web application security tests.

Figure 2.4 – Burp Suite, a tool used for web application security testing

Conducting regular penetration tests provides organizations with validation of their security controls, plus the ability to uncover hidden threats before they become too serious. It is an essential aspect of any strong cybersecurity program and ensures that systems remain resilient from attacks while preparing companies for real-world threats.

Social engineering tests

Social engineering tests are a vital tool for determining the potential vulnerabilities that stem from an organization's human-centric components. These tests simulate various social engineering attacks to evaluate the extent of employees' observance of security protocols.

The most common type of test is a phishing simulation, which involves sending malicious emails to employees to assess their ability to recognize and report attacks.

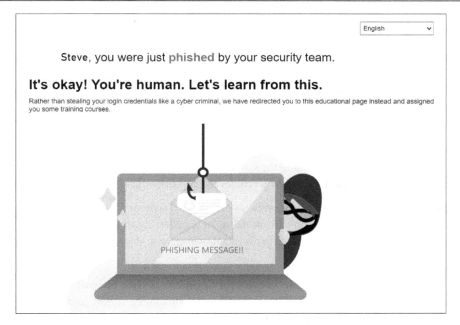

Figure 2.5 – Phishing simulation example

Other social engineering tests include pretexting tests, which occur when an attacker fabricates a false scenario to acquire confidential information or unauthorized access to systems. Impersonating an IT support person who requests a password reset is one example of such a deception.

Tailgating tests examine the effectiveness of physical security measures while also testing employees' adherence to these principles by attempting entry into restricted areas by following authorized personnel after creating some sort of urgency or relying on politeness.

Baiting tests use malicious devices, such as USB drives, as bait that curious employees may unknowingly plug into a computer and inadvertently install the malware.

The results from social engineering tests are highly beneficial to understanding how humans influence an organization's security posture. Through these assessments, areas where employees require additional training and awareness can be identified and highlighted, illustrating that strong cybersecurity is not just about technology but also people and their decisions. Such tests further emphasize the need to cultivate a security-first culture within any organization since humans are the weakest link in any cybersecurity defense strategy.

Social engineering tests are essential to any organization's security system. They play a significant role in determining the weak points of an organization's human-centric defenses and can help identify areas where further training and awareness are needed. Ultimately, these tests serve as vital tools for uncovering potential vulnerabilities that may arise from human error or negligence.

Conducting risk assessments

Organizations can protect their valuable data and infrastructure by conducting regular assessments and implementing risk mitigation strategies. Let's start by learning more deeply about risk assessment. Various risk assessment methodologies exist, such as NIST SP 800-30 and ISO 31000, which provide step-by-step guidelines for conducting comprehensive assessments:

- **NIST SP 800-30** is a risk assessment methodology developed by the **National Institute of Standards and Technology (NIST)**. It provides step-by-step guidelines for conducting assessments, including identifying assets, defining the scope, identifying threats and vulnerabilities, assessing their likelihood and impact, calculating risk levels, and prioritizing risks.

- The **ISO 31000** risk assessment process helps organizations proactively manage potential risks by offering guidance on preventing, minimizing, or transferring those risks. Organizations can ensure compliance with industry standards by following these steps in their organization-wide risk management process.

Risk assessment methodologies are also helpful for compliance with industry regulations and frameworks such as the ISO 27001, PCI-DSS, or the EU's GDPR subject to the business, industry and geo-political requirements. These regulations require organizations to comprehensively assess their security posture and take steps to mitigate any identified risks.

In simple terms, organizations can keep their data safe by regularly checking for potential risks and taking steps to lessen them. The NIST SP 800-30 and ISO 31000 methods can be considered 'how-to' guides for this process, helping identify what needs protection, determining what threats exist, and deciding how to handle these risks. These methods also allow organizations to meet industry rules, such as PCI-DSS for payment card security or GDPR for data protection in Europe, which obligate them to thoroughly check their security and fix any potential issues.

Risk assessment methodologies

Risk assessment methodologies are essential for organizations to successfully identify, assess, and mitigate cybersecurity risks. Several methodologies are available, each providing a structured framework enabling organizations to perform comprehensive evaluations:

- **Factor Analysis of Information Risk (FAIR):** FAIR is a quantitative risk assessment methodology that leverages mathematical models to compute risk probability based on threat frequency, vulnerability levels, and anticipated consequences.

- **ISO 31000:** As covered previously, ISO 31000 provides a general risk management framework applicable to various domains, including cybersecurity. This standard promotes a risk-based approach in decision making, while guiding organizations to establish proper processes and controls for risk management.

- **Hazard Identification, Risk Assessment, and Control (HIRAC):** HIRAC is mainly utilized for health and safety assessments but can be adapted for cybersecurity assessments. It involves hazard identification to identify risks associated with each hazard as well as establishing measures to control these risks.

- **Operationally Critical Threat, Asset, and Vulnerability Evaluation (OCTAVE):** OCTAVE emphasizes the engagement of stakeholders in the process and focuses on organizational risk by assessing critical assets, potential threats, and vulnerabilities, analyzing impact levels, and developing effective mitigation strategies.

To ensure the effectiveness of any chosen methodology, its implementation needs to be tailored accordingly, considering the industry-specific and organizational requirements such as regulatory compliance and available resources. With the help of an adequate risk assessment methodology applied consistently, an organization can bolster its security posture substantially, thereby mitigating possible losses due to security incidents.

Identifying assets and establishing the scope

The first step in performing a risk assessment is identifying and documenting the organization's assets. This is critical as it ensures we can establish the scope of the assessment. Simply speaking, you cannot assess what you don't know about. This step lays the foundation for a comprehensive and focused risk assessment process. Here are the key considerations involved:

- **Asset Identification:** It is vital to identify all relevant assets within the organization and document them accurately. Doing so ensures that all critical assets are adequately safeguarded.

- **Asset classification:** Asset classification, covering the given asset's risk level, importance, and value, helps an organization evaluate the risk factors of its assets and prioritize resources accordingly.

- **Asset ownership and responsibilities:** Clearly define the responsibilities of asset owners and IT teams in asset management and security.

- **Data flow analysis:** This involves pinpointing significant data repositories, access points, and data transfer methods. Such an analysis allows for a better understanding of any issues that could harm the confidentiality or integrity of the data.

- **Boundary definition:** Establish the boundaries of the evaluation, including specifying which networks, systems, departments, geographical locations, and third-party associations are included in the assessment.

- **Regulatory and compliance considerations:** All processes should be designed to monitor and address any changes in regulations and standards to maintain compliance.

Organizations can boost their cybersecurity resilience by accurately pinpointing assets and defining the scope of their risk assessment. Such an approach helps streamline the application of resources, conduct a more thorough risk analysis, and develop effective risk mitigation strategies. It also ensures that the risk assessment process is tailored to the organization's objectives and goals, enabling more effective risk management and increased security.

Identifying threats and vulnerabilities

Organizations must take steps to identify potential threats and vulnerabilities that could compromise the confidentiality, integrity, or availability of their information systems. To do this, they must systematically evaluate internal and external factors that could pose risks to their assets.

Threat identification involves recognizing potential threats – both deliberate and accidental – that could exploit vulnerabilities and have a negative effect on assets. This includes identifying internal threats, such as insider threats or employee negligence, and external threats, such as hackers, malware, or physical breaches. Organizations must also stay informed about the latest security trends and attack vectors by leveraging threat intelligence sources, such as cybersecurity news and industry reports.

Vulnerability assessments should be conducted to uncover weaknesses or gaps in the organization's systems, software, or processes that threats might exploit. This can include automated scans or manual reviews. Additionally, organizations should practice patch management by regularly monitoring vendor updates and applying the necessary patches and updates to minimize the possibility of known vulnerabilities being exploited. Configuration management is also essential for reviewing system configurations to ensure they are secure and best practices are being followed. Misconfigured systems can introduce new holes that attackers can exploit.

Physical security assessments should also be conducted to evaluate access controls, surveillance systems, environmental protections, and other physical security controls that could be vulnerable to exploitation if not appropriately secured. Furthermore, any third-party vendors or partners with access to the organization's systems or data should be evaluated for their security practices before granting access. Third-party risks should not be taken lightly since any weaknesses they introduce could have significant consequences.

By comprehensively assessing threat and vulnerability levels, organizations can better understand what kind of risks they face and prioritize their mitigation efforts accordingly. Regularly updating these assessments is essential for keeping up with emerging trends in the cybersecurity landscape so that organizations continue protecting their assets effectively over time.

Assessing likelihood and impact

One of the critical aspects of a risk assessment process is assessing the likelihood and potential impact of threats and vulnerabilities to an organization. Essentially, you are looking to understand the probability of an attack path. By doing so, you can prioritize risk and allocate resources more effectively. The following are several critical factors to be aware of:

- **Impact assessment**: Organizations must consider the potential consequences of the identified risk being successfully exploited, including the effects on operations, assets, financial reputation, compliance, and customer and partner trust. The impact can be categorized as low, moderate, high, or quantified in terms of financial loss or system downtime.

- **Risk scoring**: Risk scores are assigned by combining the likelihood and impact assessments through qualitative rating systems or mathematical formulas/risk matrixes, which help prioritize risks based on criticality and the potential impact on the organization.

- **Likelihood assessment**: This involves analyzing threat intelligence data, historical data, and industry trends to assess the probability of a threat exploiting a vulnerability. Likelihood can be expressed qualitatively (e.g., low, medium, or high) or quantitatively (e.g., as a percentage or frequency).

- **Subject matter expertise**: SMEs with in-depth domain knowledge of an organization's systems processes and industry should be consulted for their expertise, which can add value to the assessments.

- **Data gathering**: To assess likelihood and impact accurately, organizations should gather relevant information from various sources, including internal data such as incident reports and system logs, and external sources such as industry reports and benchmarking data.

- **Documentation**: Assessments must be documented with clear explanations and justifications for ratings/scores provided for each risk identified to provide reference material for decision-makers and facilitate communication throughout the organization about risks encountered. There should be no room for interpretation or guessing what something could mean.

Carrying out regular reviews and updates of assessments helps keep them aligned with the evolving threat landscape and organizational context, ensuring organizations remain adequately protected against risks. Accurate assessments of risk likelihood and impact enable organizations to focus resources on higher-risk areas and develop effective mitigation strategies, improving their overall security posture and resilience against malicious actors threatening their business objectives.

Prioritizing risks and developing mitigation strategies

After assessing the likelihood and impact of identified risks, the next step in this process is calculating risk levels for each risk. Risk levels provide a quantitative or qualitative representation of the overall risk associated with each threat-vulnerability pair.

Organizations should first establish criteria that define thresholds for different risk levels to calculate these levels accurately. These criteria should be based on organizational policies, industry standards, or regulatory requirements to ensure consistency and facilitate decision-making regarding risk management strategies. A risk matrix or scoring model can then be utilized to calculate these levels by mapping the assessed likelihood and impact to its corresponding cells in the matrix grid, which will assign a specific numerical value or rating (e.g., low, medium, or high) to each risk based on the calculated values. In some cases, a quantitative approach can also assign numerical values to likelihood and impact via setting probability values, estimating potential monetary losses, or utilizing mathematical formulas to calculate scores.

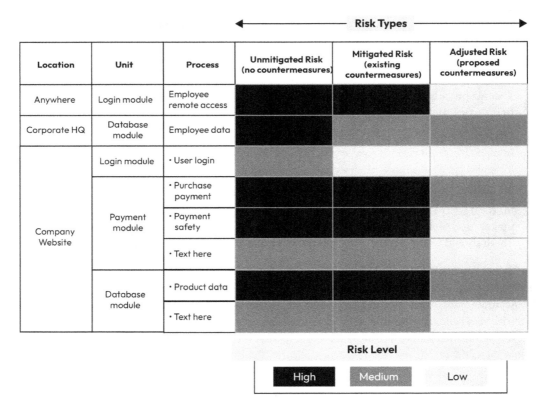

Figure 2.6 – Risk assessment example template

Once calculated, it's essential to communicate the calculated risk levels to stakeholders, decision-makers, and relevant teams, as this ensures a shared understanding of the risks and enables informed decision-making regarding risk treatment options. Additionally, documentation of these calculations must be kept for future assessments and as a historical record of the organization's risk scale.

It's essential that all organizations continuously review and update their calculated risk levels to adapt to any changes in their threat landscape and their organizational context or risk tolerance. By doing so, they'll understand their risks better, allowing them to prioritize effective mitigation efforts accordingly.

To summarize, after identifying potential risks, every organization needs to determine the level of each risk, quantitatively or qualitatively. This involves setting criteria based on company policies, industry norms, or legal requirements, and then using a risk matrix or scoring model to assign a rating to each risk. It's crucial to share these ratings with everyone involved in decision-making to ensure everyone is on the same page about the risks at hand. Keep a record of these calculations for future reference and make sure to update them regularly to account for changes in the organization or its risk tolerance. This will help the company better understand its risks and decide where to focus its efforts to prevent them.

Documentation and reporting

Effective risk assessment requires comprehensive documentation and reporting to capture the process's findings, methodology, and outcomes. To ensure consistency and clarity in the documentation, organizations should develop standards and templates for documenting the purpose, scope, objectives, methodologies, and tools used for risk assessment. In addition to creating documents that provide an overview of the process, organizations must keep detailed records of all identified risks with their likelihood and impact assessments, risk levels, and supporting data or calculations. This information will be invaluable for future assessments, allowing organizations to track changes over time.

Organizations should also create a risk register or repository to store all the identified risks along with essential information such as the description of the risk, risk levels, priority rankings, and proposed mitigation strategies. This centralized repository allows organizations easy access to risk information while enabling them to manage ongoing risk management efforts more efficiently.

Risk assessment reports are vital to effective risk assessment as they provide stakeholders with a clear overview of the assessment's findings and recommendations. These reports should include an executive summary describing the purpose of the assessment:

- A methodology overview
- Description of assets and scope
- Identified risks with their likelihood and impact assessments
- Corresponding risk levels
- Actionable insights on mitigating these risks

- Visual representation (such as charts/graphs) for better understanding
- The timeline/responsibilities/resources required for implementing recommended mitigation strategies

The risk assessment report should be tailored appropriately based on audience, ensuring that it provides enough detail suitable to each stakeholder, be it management, decision makers, or IT teams, while also using non-technical language for ease of understanding by everyone who reads it. Only authorized individuals/teams should be given access to sensitive information after ensuring proper access controls and data protection protocols are in place. Furthermore, a record retention policy should be established to comply with legal and industry requirements while considering organizational needs.

By providing accurate documents and transparent reports, stakeholders can get an insight into various aspects, such as the risks identified, their potential impacts, and the suggested mitigation strategies, which helps facilitate informed decision-making and enables prioritization efforts while providing a foundation for ongoing risk management and cybersecurity initiatives. Documents and reports must be regularly updated and reviewed to reflect changes in the landscape and maintain accuracy throughout all process stages.

Monitoring and reviewing

Organizations must monitor and review the risk assessment process to ensure their assessments remain up to date, relevant, and aligned with the evolving threat landscape and organizational context. Always remember that in the cybersecurity world, everything continuously evolves rapidly; therefore, we cannot rely on one-off risk assessments, nor can we assume that risk assessment processes built years ago still apply or provide the same level of benefits.

Establishing regular mechanisms for monitoring the risk landscape through technological changes, industry trends, regulations, emerging threats, security incidents, or breaches within the organization or industry can help identify new or emerging risks that were not previously considered. Regular reviews must be conducted to determine whether updates are necessary according to factors such as the organization's risk appetite, industry best practices, regulatory requirements, or significant changes in the business environment.

To ensure the effectiveness of the process, it is important to involve stakeholders such as IT teams, management, and SMEs in assessing and refining risk management practices. This helps gain perspectives and feedback on any ongoing adjustments or updates. Additionally, **key performance indicators (KPIs)** related to risk reduction, incident response, or security controls should be monitored to evaluate their impact on risk levels and overall security posture. Upon making any changes, it is also essential to document these changes to maintain a clear audit trail of historical records.

Finally, it is also crucial to communicate any significant updates or changes with decision-makers by providing them with updated risk assessment reports that should highlight vital changes and explain the rationale behind adjustments in risk prioritization or mitigation strategies.

Businesses need to keep a constant eye on their cybersecurity practices. The digital world changes rapidly, and threats to security continue to increase and rapidly evolve, so we can't rest on our laurels. By regularly checking in on the state of things, involving the right people in discussions, and keeping track of key performance indicators, we can stay on top of potential risks. It's also essential to keep a record of any changes made and keep important people in the loop with comprehensive reports on any changes and the reasons behind them.

Prioritizing and remediating weaknesses

The process of prioritizing and remediating weaknesses is essential and involves assessing the associated risks, evaluating their potential impacts, and determining the severity of the vulnerabilities they expose. By understanding the risk landscape, organizations can better focus on vulnerabilities that pose the greatest threat to their operations and data.

Remediating weak spots requires a systematic approach considering risk levels, cost-effectiveness, compliance obligations, and the organization's specific threat landscape. It also involves collaboration between cybersecurity teams, management, and other stakeholders to ensure alignment and support for remediation efforts. Additionally, proper documentation and reporting are integral for providing transparency, accountability, and a historical record of weaknesses.

To help mitigate identified vulnerabilities, targeted strategies should be implemented, such as applying patches, updating software, configuring secure access controls, implementing security controls, and enhancing employee training and awareness. The goal is to minimize the attack surface and reduce the likelihood of successful exploitation. Furthermore, continuous monitoring of emerging threats and evolving vulnerabilities through regular vulnerability scanning, penetration testing, or incident monitoring will help identify new weaknesses that must be prioritized and remediated.

Tools such as **Threat and Vulnerability Management (TVM)**, **Security Posture Management (SPM)**, and **Attack Surface Management (ASM)** can aid security teams in these aspects. Next-generation technologies in these spaces are geared toward continuously discovering misconfiguration, vulnerabilities, and threats and helping prioritize them by the potential impact on the organization. By dedicating resources to address vulnerabilities strategically through effective prioritization and remediation processes, organizations can significantly reduce the risk of security breaches while protecting critical assets with sensitive data and upholding customer trustworthiness.

Understanding risk and impact levels

Organizations must assess the risk levels and potential impacts of identified cybersecurity weaknesses to prioritize and remediate them effectively. Risk assessment involves evaluating the severity of the vulnerabilities, the likelihood of exploitation, and the possible consequences of a successful attack. Impact assessment requires organizations to consider the potential damage to critical systems, loss of sensitive data, operational disruptions, financial losses, reputational harm, and compliance with regulations.

Organizations can understand the risks of weaknesses by holistically analyzing risk levels and impact assessments. This understanding allows informed decision-making on which weaknesses should be prioritized for remediation. High-risk vulnerabilities with significant potential impact should be given immediate attention, while weaknesses with lower risk levels or minimal potential impact may receive attention during subsequent remediation cycles.

With this approach, organizations can focus their efforts on vulnerabilities that pose the greatest threat to their security posture and business continuity. Proactive remediation of high-risk weaknesses will help reduce the organization's attack surface and enhance its overall cybersecurity resilience.

Risk mitigation strategies

Risk mitigation strategies are designed to reduce the likelihood and impact of potential risks by introducing specific measures to address weaknesses and vulnerabilities within the cybersecurity framework. The most critical aspects when it comes to risk mitigation strategies include the following:

- **Patch management** is one of the most common risk mitigation strategies, and involves regularly applying security patches and updates released by software vendors to address known operating system, application, and firmware vulnerabilities.

- **Access control measures**, such as robust authentication mechanisms, the principle of least privilege, multi-factor authentication, and strict password policies, are also essential for protecting against cyber threats.

- **Network segmentation** is an effective strategy for isolating networks into smaller components to limit the spread of any potential security breach or unauthorized access.

- Organizations should implement **data encryption** for both in-transit and at-rest data using end-to-end encryption for communications and full-disk encryption for storage devices.

- **Cybersecurity awareness training** programs can help educate employees about identifying potential threats, safe online practices, and reporting incidents promptly.

- **Develop incident response plans** outlining incident escalation procedures and conduct regular drills to respond to security events efficiently.

- **Security monitoring and logging** is another critical component of a comprehensive risk mitigation plan – organizations should use **Extended Detection Response (XDR)** and **Endpoint Detection Response (EDR)** solutions to detect suspicious activities or cyber threats quickly.

- It is imperative that organizations also **assess and manage third-party risk** from vendors or service providers given access to their systems or data. This can be done by conducting thorough security assessments and reviewing vendors' security practices while enforcing contractual security requirements.

- **Backup and disaster recovery plans** are vital for business continuity in case of a system failure or security incident. Organizations should regularly back up critical data to offsite locations while testing the restoration process periodically.

- **Using secure coding practices** during software development can mitigate the introduction of vulnerabilities. These practices include code reviews, the use of secure coding frameworks, and employing **secure development lifecycle** (**SDL**) methodologies.

Organizations can strengthen their cybersecurity posture by taking the necessary steps to implement these risk mitigation strategies tailored to their specific needs and requirements, such as industry regulations and the broader threat landscape.

Attack surface reduction

Organizations can significantly reduce their risk of cyber threats by focusing on attack surface reduction. By implementing the proper security measures, organizations can strengthen their overall security posture and safeguard critical assets from unauthorized access or compromise. Here are the essential considerations for attack surface reduction:

- **Inventory and asset management**: Gain a complete understanding of the organization's digital assets, including software, applications, hardware, and data repositories.

- **Incident response planning**: Develop and implement an incident response plan for effective response and mitigation in the event of a security incident. Establish an incident response team, define escalation procedures, and regularly conduct drills to ensure readiness.

- **Patching and vulnerability management**: Ensure operating systems and applications are kept up to date with the latest security patches and updates. Establishing a robust vulnerability management program, including continuous vulnerability scanning, prioritizing patching efforts, and the timely application of patches to mitigate known vulnerabilities, is critical.

- **Network segmentation**: Divide networks into smaller, isolated segments to limit the lateral movement of attackers. Implement network segmentation, firewalls, and access controls to restrict communication between network segments and determine the potential impact of a security breach.

- **Regular audits and assessments**: Perform security audits, vulnerability assessments, and penetration testing to identify weaknesses and gaps in the organization's security controls. Regular assessments help to detect new attack vectors and address evolving threats.

- **Least-privilege principle**: Apply the principle of least privilege by granting users only the minimum level of access necessary for their tasks.

- **User awareness and training**: Educate employees about potential security risks and safe online practices. Regularly conduct security awareness training to promote a security-conscious culture and empower employees to recognize and report threats.

- **Third-party management**: Assess the security practices of third-party vendors with access to our systems or data. Conduct due diligence in selecting and vetting third parties, and enforce solid contractual obligations for security controls and incident response protocols.

- **Continuous monitoring**: Implement robust monitoring mechanisms to detect and respond to security incidents. Monitor network traffic, system logs, and user activities to identify and investigate malicious and suspicious behavior.

- **Secure configuration**: Configure systems, applications, and devices with security in mind. Disable unnecessary services and protocols. Implement secure communication protocols. Enforce strong password policies, and ensure your organization follows the industry standards and the best practices from vendors.

- **Application security**: Ensure secure coding practices are followed during application development. Techniques for this include the use of code reviews, input validation, secure coding frameworks, and regular security testing such as **Static Application Security Testing (SAST)** and **Dynamic Application Security Testing (DAST)**.

- **Cloud security**: Put appropriate security measures and configurations into place for cloud environments. Employ strong access controls, encryption, and monitoring tools to protect data and applications hosted in the cloud.

- **System hardening**: Implement security best practices and guidelines to harden systems and devices. This includes configuring systems to disable default accounts and passwords, removing unnecessary services, and applying secure configuration baselines provided by vendors or industry frameworks.

- **Physical security**: Implement security measures to protect critical infrastructure, servers, and data centers. This includes controlling access to sensitive areas, installing surveillance systems, and monitoring physical access points.

Organizations can reduce their exposure to potential cyber threats by focusing on attack surface reduction. By implementing these measures, organizations can significantly enhance their security posture, improve their ability to detect and respond to attacks, plus safeguard critical assets and sensitive data from unauthorized access or compromise. Adopting a proactive approach of continuously monitoring and adapting these attack surface reduction strategies is essential for addressing emerging threats and evolving technologies.

Continuous monitoring and reassessment

Organizations can improve their cybersecurity resilience by embracing continuous monitoring and regular reassessment. Through advanced technologies such as EDR and XDR systems, organizations can detect and respond to security incidents in real time, allowing threats to be contained and cyber incidents to be responded to faster.

Implementing EDR and XDR solutions enables real-time threat detection, granting organizations visibility into all assets so that malicious activities can be identified quickly and effectively. Additionally, these systems allow automated workflows for threat detection, alerting, and remediation, dramatically reducing response times and increasing operational efficiency.

Continuous monitoring also facilitates proactive threat-hunting activities where security teams actively search for indicators of compromise or emerging threats. Furthermore, behavioral analysis techniques are used to identify anomalous behavior or deviations from standard patterns, which helps detect sophisticated attacks often missed by traditional security controls. User activity monitoring is also critical to identify potential insider threats or unauthorized access, while network traffic monitoring helps uncover malicious connections or data exfiltration attempts. Centralized log management and analysis tools also play a role in continuous monitoring as they provide visibility into system logs, helping identify security events or indicators of compromise. Organizations can further extend their capabilities in detecting emerging threats by leveraging real-time threat intelligence feeds.

Lastly, when it comes to cloud environments, continuous monitoring must extend here, too, as cloud security monitoring tools help monitor the security posture of cloud infrastructure, applications, and data. It is also essential that regular reassessments take place for organizations to stay on top of vulnerabilities or gaps within their systems that attackers could exploit. This enables organizations to adjust their security controls according to the ever-evolving threat landscape and generate audit logs required for regulatory compliance assessments.

Summary

To protect against cyber threats, organizations must take a proactive approach. This includes identifying all digital assets, setting up a response plan for potential security incidents, keeping systems updated with the latest patches, and dividing networks into manageable segments. Regularly auditing and assessing security controls, only granting necessary access privileges, training staff on security risks, and managing the risk posed by third-party vendors are also crucial. Additionally, it's essential to monitor network traffic, configure systems securely, follow safe coding practices, employ security measures for cloud environments, and physically secure infrastructure.

More so, companies need to continuously monitor their systems and reassess their strategies regularly. With technologies such as EDR and XDR, real-time threat detection is possible, allowing organizations to identify and respond to malicious activities promptly. They also need to implement proactive threat-hunting activities and use behavioral analysis to detect unusual activities. Monitoring user activity, network traffic, and managing and analyzing system logs further help identify potential internal threats and malicious connections. It's equally important to ensure that these practices extend to cloud services. Regular reassessment allows companies to identify and fix system vulnerabilities and meet regulatory requirements. In the next chapter, you will learn how you can monitor for emerging threats and trends to stay ahead of the curve.

3
Staying Ahead: Monitoring Emerging Threats and Trends

There is a relentless struggle between organizations striving to safeguard their digital assets and threat actors who continually devise sophisticated methods to breach these defenses. At some point, the question always arises: how do we turn the battle around? And how can we predict the next move of an adversary? The cybersecurity landscape will always be rapidly changing; therefore, the real question is: how can we become more vigilant, gain a deeper understanding of threat actors to predict their moves, and leverage technological innovation to our advantage? As security professionals and teams, we must transition from reactively responding to cyber threats and become proactive. Therefore, we must stay informed of emerging threats and trends by constantly monitoring threat intelligence feeds, cybersecurity news, and professional and personal networks, and simply develop effective countermeasures. It's important to remember that it's not if but when our organization will be targeted. We must adopt a threat actor's mindset and with that, understand their motivations, strategies, and capabilities. We must constantly challenge ourselves and the status quo, innovate, and not limit ourselves to technology, but also extend to fostering a culture of continuous learning. Adopting emerging technologies such as **artificial intelligence (AI)**, **machine learning** (**ML**), and zero-trust models can bolster defenses, but ultimately, staying ahead of evolving threats is a constant game of cat and mouse, requiring adaptability, creativity, and resilience. We will delve into the significance of keeping abreast of emerging threats in the cybersecurity landscape, equipping you with the knowledge needed to anticipate and counteract potential infiltration attempts. You will gain insight into the attacker's mindset and understand their tactics, strategies, and motivations, which is critical for effective defense. Moreover, we will explore the pivotal role of innovation in cybersecurity, emphasizing how continuous technological advancements can bolster defenses, ensuring a robust and resilient security posture against evolving threats.

This chapter will cover the following topics:

- The importance of monitoring emerging threats and trends
- The attacker's mindset
- The role of innovation in cybersecurity

Let us dive in!

The importance of monitoring emerging threats and trends

When considering monitoring emerging threats and trends, you can consider it like having a radar system. Similar to how radar can detect incoming objects and provide real-time information to guide a ship through challenging waters, monitoring emerging threats and trends allows organizations to detect and anticipate potential cyber threats. This means it can act as a protective shield, enabling proactive response and mitigating risks by providing early warning signals and valuable insights.

It is paramount to monitor for emerging threats and trends and to be successful in this, we must first understand the evolving threat landscape. Rest assured that just like, as defenders, we gain access to new technologies, so do our adversaries. In fact, most often, you will find adversaries being able to utilize new technologies faster than defenders as they are not bound to compliance, legal, or organizational boundaries that may slow down innovation. With threat actors consistently innovating, new threats such as deep fakes, AI-powered attacks, and ransomware-as-a-service pose significant risks to organizations. Closing your eyes against these threats is not an option. It is crucial to stay informed about these threats through reliable threat intelligence sources, including cybersecurity news, feeds, bulletins, and professional networks. This awareness is not just to understand the risks but also to convert this knowledge into proactive, actionable strategies. By staying vigilant and responsive to these emerging trends, organizations can update their policies, implement appropriate defenses, and train staff effectively, ensuring robust and timely protection against the ever-changing threat landscape.

Understanding the cybersecurity landscape

Understanding this dynamic environment we are in within cybersecurity requires consistent analysis of the adversaries involved, their motivations and tactics, technological trends, and potential defensive measures. The threat landscape consists of cybercriminals motivated by financial gain, sophisticated criminal groups, and nation-states engaged in cyber warfare and espionage. Each adversary presents unique challenges based on their methods, targets, objectives, and available resources. From **advanced persistent threat** (**APT**) groups focused on ransomware attacks to APT groups used for long-term surveillance by state-backed entities – the strategies are often drastically different from one another. To navigate successfully through the complex environment of cyberspace, it is essential to comprehend all its nuances.

Cyber threats are growing in complexity as traditional attacks such as malware, phishing, and basic forms of social engineering are increasingly becoming obsolete and replaced by more sophisticated threats including AI-driven attacks, deep fake technology, and sophisticated ransomware such as human-operated ransomware campaigns. At the same time, technology provides both challenges and opportunities – from the proliferation of IoT devices to cloud computing to 5G networks – that expand attack surfaces for cyber adversaries while offering new defense mechanisms against them. In the evolving cybersecurity landscape, threats have become increasingly sophisticated with the advent of technology such as AI, deepfake technology, and cloud computing. These technologies have been leveraged in attacks such as the "DeepLocker" attack, where AI was used to hide malware within benign applications, or instances such as the case where a CEO was fooled by a deepfake voice into transferring funds. Human-operated ransomware campaigns such as "Sodinokibi" target vulnerable companies with little IT security, while the proliferation of IoT devices escalates the threat of Mirai botnet attacks. Furthermore, cloud computing has opened up new threat vectors, as demonstrated by the Capital One breach where a misconfigured web application firewall was exploited. Lastly, the rise of 5G networks introduces unique cyber threats, with potential misuse of network slicing features allowing hackers to bypass traditional security measures.

Organizations must monitor the latest cybersecurity technology and methodology developments to stay ahead of these threats. New technologies such as **extended detection and response (XDR)** and threat-hunting methods can detect and respond to potential threats faster than ever before. Additionally, keeping up with emerging governmental norms and standards for cybersecurity is essential for meeting legal requirements and strengthening defenses.

Threat-hunting methods are proactive cybersecurity strategies used to detect malicious activities within a network that may evade existing security systems. There are several recognized methods in this discipline:

1. **Hypothesis-based threat hunting**: This method starts with a hypothetical situation, such as a specific attack scenario, and then hunts for evidence that supports or refutes it.

2. **Indicator-of-compromise (IoC) based hunting**: This method involves searching for known indicators of a breach, such as specific types of malware, suspicious IP addresses, or unusual network traffic patterns.

3. **TTP- (tactics, techniques, and procedures) based hunting**: This method focuses on understanding the behavior of attackers. The process involves identifying and detecting the tactics, techniques, and procedures used by cybercriminals, which are often unique to specific threat groups.

4. **ML and automated threat hunting**: This method employs algorithms to identify abnormalities or anomalies that may signify a security breach. These anomalies could include unexpected access to sensitive data, irregular user behavior, or unusual network activities.

Each of these methods prioritizes proactive measures to identify threats, focusing on detection, analysis, and mitigation before significant damage occurs.

Organizations must be proactive in their approach toward cybersecurity in order to stay ahead of their adversaries. To do this, they must stay up to date with the latest technologies, policies, and methodologies related to cybersecurity, ensure that their staff is adequately trained on cyber safety protocols, and keep an eye out for unique opportunities for innovation. Taking these measures will enable them to remain resilient against a wide range of malicious actors. Additionally, organizations should strive to continuously monitor governmental norms and standards around cybersecurity and invest in the necessary resources needed to bolster their defenses against emerging threats. By adopting a comprehensive strategy that focuses on both prevention and preparedness, organizations can position themselves for success when it comes to protecting their digital assets.

The risks of emerging threats

The ever-evolving nature of cyber threats has been demonstrated through numerous real-world examples, indicating that such risks are far from hypothetical and can have severe implications for businesses and society. AI-powered attacks, for instance, have been used to replicate writing styles and convince senior executives to send large sums of money.

An example of the potential malicious use of deepfakes was recently seen in a case involving a United Kingdom energy company. The CEO was deceived into transferring a large sum of money – an amount totaling $243,000 – after being tricked by a deepfake video. This instance shows how deepfakes can be applied to commit financial fraud and highlights the need for greater vigilance in verifying sources of information.

In 2016, the Mirai botnet revealed just how dangerous unsecured **Internet of Things (IoT)** devices can be when it conducted large-scale **distributed denial-of-service (DDoS)** attacks on major websites. The Mirai botnet served for many in the industry as a wake-up call and pushed us as an industry to understand the importance of securing vulnerable IoT infrastructure.

The Capital One data breach of 2019 demonstrated that security teams can no longer ignore cloud resources, including virtual machines and applications. The Capital One data breach resulted in many organizations starting to look into how to shift left and implement DevSecOps frameworks to improve the security state of their cloud infrastructure.

The potential for significant financial loss, reputational damage, and regulatory penalties further underscores the importance of protecting against emerging cyber threats. The British Airways data breach in 2018, which resulted in a £183 million fine under the **General Data Protection Regulation (GDPR)**, demonstrates why these risks must be taken seriously. With advanced security tools such as **endpoint detection and response (EDR)**, **threat and vulnerability management (TVM)**, and regular system updates, organizations can better mitigate the risks associated with emerging cybersecurity threats. By adopting a proactive approach to cybersecurity and protecting their systems from today's and tomorrow's threats, businesses can ensure they are well-equipped to face whatever challenges come their way.

In conclusion, real-world examples of emerging cyber threats illustrate the potential for severe operational disruption, financial loss, reputational damage, and regulatory penalties. Investing in a proactive approach to cybersecurity is essential for businesses to mitigate these risks. Advanced security tools such as XDR or security AI platforms can help detect malicious activities and enable organizations to respond quickly and effectively. Regular system updates ensure that systems stay up-to-date with the latest cybersecurity defenses. Organizations can safeguard their data, operations, finances, and reputation by taking a proactive stance against emerging threats and protecting themselves from today's and tomorrow's challenges. And though no one can predict the future of cybercrime perfectly, following these steps will go a long way in helping businesses remain secure.

The role of threat intelligence

Threat intelligence is no longer optional or only relevant for the largest organizations in the world. Threat intelligence is essential to any organization's cybersecurity efforts, providing vital insight into incident prevention and response. It encompasses various types of information, such as **indicators of attack (IoAs)**, IoCs, TTPs used by adversaries, and information about specific threat actors, including primary motivations, attack paths, and more. This data can be automatically or manually sourced from various sources such as cybersecurity news reports, public and private threat intelligence feeds, security bulletins, and others. With a comprehensive understanding of the threat landscape enabled by threat intelligence, organizations can take proactive steps to defend against potential threats and minimize the risk of a successful cyber attack. *Figure 3.1* is an example of how threat intelligence can be incorporated into a security AI platform:

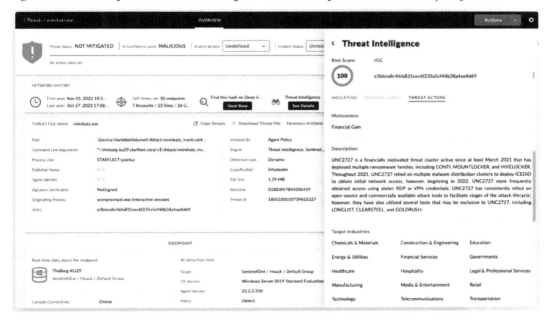

Figure 3.1 – SentinelOne Singularity Threat Intelligence

Real-world examples highlight the importance of having reliable threat intelligence. For instance, in 2013, Target experienced a data breach due to attackers stealing credit card information from 40 million customers; however, had they been equipped with sufficient threat intelligence to properly contextualize an alert that had flagged suspicious activity before the attack, the severity of the risk posed may have been recognized and prevented. Similarly, in 2014, Sony Pictures was hacked by the North Korea-linked Lazarus Group using malware called Destover; if investigators had access to enough threat intelligence to make the connection between the malware used and this group, they might have been able to attribute it much more quickly. *Figure 3.2* visualizes the kill chain of the Sony attack:

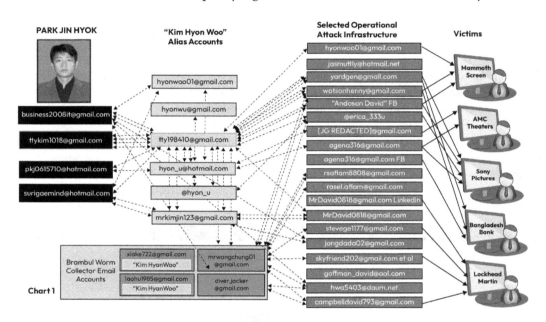

Figure 3.2 – Sony Attack kill chain

In conclusion, with ever-evolving threats, organizations must stay informed on what is happening in their environments to maintain effective defenses against cyber-attacks – this is where threat intelligence plays an invaluable role. By providing actionable insights into IoCs, TTPs used by malicious actors, and information about specific groups responsible for attacks, organizations can become proactive rather than reactive and prioritize their defensive strategies based on their unique circumstances and requirements.

From awareness to action

In the digital age, cybersecurity is not an option but a necessity. With the increasing sophistication and prevalence of cyber threats, organizations must be proactive in their defense strategies. Effective cybersecurity measures include proactive threat assessment and technical defenses, employee training, regular testing, and lessons from real-world incidents.

Proactive threat assessment

Organizations need to adopt a forward-looking approach to threat detection. Understanding and staying informed about potential threats is integral to an effective cybersecurity strategy. This involves gathering comprehensive data on threats and using this information to guide the strategic planning process, including task delegation, devising an incident response plan, and identifying areas of continuous improvement.

Implementation of technical defenses

The application of technical solutions is fundamental in protecting against known cybersecurity threats. These might include using detection and response systems to isolate and counter cyber threats on various levels, such as identity, endpoint, and cloud. Additionally, staying informed about upcoming threats can prompt the deployment of avant-garde solutions such as AI-based defense systems.

Employee training and security awareness

Awareness of security among employees is crucial to maintaining a solid defense against cyber threats. Regular and extensive training sessions should be provided to educate personnel about prevalent threats and the appropriate responses to potential incidents. Learning resources should be made available to staff, covering topics such as safe browsing, secure email usage, effective password management, and the latest trends in threats.

Regular testing and vulnerability scanning

Frequent vulnerability scanning and penetration tests help discover and address security vulnerabilities that attackers could exploit. Vulnerability scanning can identify known system weaknesses, including unpatched software and misconfigured network devices. Penetration tests can simulate real-world attacks, providing valuable insights into system robustness and uncovering additional vulnerabilities.

Valuable lessons from real-world incidents

Real-world case studies are significant reminders of the need for timely action upon identifying potential risks. The Equifax data breach of 2017, the rapid response to the Heartbleed bug, and the NotPetya attack on Maersk underscore the need for swift action and the importance of investing in robust network defense systems.

The key takeaway is that organizations must proactively implement defensive measures against potential cyber threats. This entails strategic planning, the deployment of sophisticated technical defenses, regular employee training, and routine testing. These actions are necessary to keep defenses current and capable in the face of the ever-evolving global landscape of cyber threats.

The attacker's mindset

Understanding cyber attackers' motivations, strategies, and psychological traits is essential for modern cybersecurity. Defenders are often caught thinking in checklists, meaning they checkmark security controls and assume the organization will be secure. In contrast, attackers have an advantage as they look at it from a graph perspective and pinpoint the weakest link. By taking the view of those who seek to exploit weaknesses and compromise systems, organizations can fortify their defenses and proactively ward off potential threats. This section will explore the benefits of adopting the mindset of an attacker, from increasing threat awareness to developing better countermeasures. It will also discuss the ethical considerations and legal boundaries that must be considered when attempting to protect one's data and systems. By delving into the mindset of cyber attackers, security professionals can strengthen their cyber defenses and better anticipate evolving digital threats.

The significance of understanding the attacker's perspective

Understanding the attacker's perspective is of paramount importance. The more aware you are of the attacker, the better your chances to prevent, detect, and respond to cyber threats. Organizations can bolster their defensive measures and proactively protect their data and systems by gaining insight into attackers' motivations, tactics, and tools.

Adopting the attacker's point of view allows companies to anticipate potential threats and identify weaknesses that attackers may target. This enables them to develop sophisticated defense strategies that specifically target these areas of vulnerability. Moreover, this helps organizations prioritize their security efforts by recognizing and responding to the objectives and goals of attackers.

Furthermore, by putting themselves in a hacker's shoes, security professionals can stay ahead of emerging threats and attack vectors that are continuously evolving. If you want to put yourself in the hacker's shoes, one of the most important steps is understanding how hackers think and act. Start by gathering information about their techniques and strategies for attack. Read blogs, newsletters, and books about hacking tactics, defenses, and reverse engineering. Study flowcharts of malicious code architectures or watch white-hat hacking demonstrations from trusted sources. Additionally, consider learning primary programming languages such as HTML and JavaScript to help you better understand what a hacker can do with code. Lastly, gain hands-on experience via online coding challenges or virtual machines that simulate cyber threats; this will not only test your skills but also give you an understanding of how to defend against potential hacks. With these steps in mind, you can begin to think more critically like a hacker and gain the necessary skillset they use daily in their work!

Additionally, this perspective allows for better identification of IoCs when responding to incidents and faster detection and containment of breaches. Understanding the attacker's mindset also leads to improved communication between security teams and other stakeholders, facilitating collaboration regarding risks and security measures.

Overall, understanding the attacker's viewpoint is essential for establishing a robust, proactive cybersecurity posture that can keep up with the quickly changing threat landscape. By taking on an adversary's perspective, security professionals can better anticipate upcoming threats while providing enhanced incident response capabilities to swiftly respond to any security incidents with maximum effectiveness.

Motivations and objectives of attackers

Cyberattacks conducted by malicious actors can have various objectives and motivations, with financial gain often being the top priority. Attackers looking to monetize their activities may resort to stealing sensitive data or launching ransomware or fraud campaigns. Other attackers may be motivated by espionage, targeting organizations that possess valuable intellectual property or confidential information in an effort to acquire trade secrets, research data, or classified information. Additionally, some attackers may attempt to disrupt and even destroy public services and critical infrastructure in order to cause extensive chaos and economic harm. Hacktivism is another form of attack used to promote social or political causes through website defacement, DDoS attacks, or data leaks. Some individuals may also hack for thrills, using it as a way to challenge themselves and showcase their technical abilities.

Real-world examples of malicious actors launching various attacks for various motives can be seen in many incidents. In 2018, ransomware attackers targeted multiple hospitals in the United Kingdom, demanding payment to regain access to systems and patient data. The attack resulted in the diversion of patients and disruption of medical services, resulting in an estimated financial loss of over £100 million. Financial gain is also a common motive for cybercriminals who use phishing campaigns or malware to steal credit card information or financial data from unsuspecting victims.

Espionage-motivated attackers have also been responsible for several high-profile incidents, such as the Office of Personnel Management data breach that affected over 22 million employees and contractors, or the Stuxnet attack on Iranian uranium enrichment facilities, which reportedly caused severe damage to its nuclear program. Disruption and destruction attacks often target critical infrastructure, such as power grids or transportation networks, to cause economic harm or civil unrest. For instance, a large-scale cyberattack against Ukraine in 2015 crippled government websites, banks, energy plants, and media outlets for over a week with significant economic losses reported.

Hacktivism is another form of attack used to advocate for social or political causes through defacing websites or leaking confidential information. Notable cases include Anonymous's attacks on companies such as PayPal and MasterCard during the Occupy Wall Street movement to protest their refusal to process payments for WikiLeaks donations. Thrill-seeking attackers may seek out vulnerabilities for fun without any financial gain or sociopolitical agenda involved. Such activities can result in severe consequences if not correctly handled by experienced professionals due to their potential implications on security and privacy.

Organizations must understand the various objectives of malicious actors to properly defend against them. To create an effective cybersecurity strategy, organizations should prioritize security measures and implement countermeasures based on specific risks. Having an in-depth understanding of what attackers are trying to achieve allows organizations to take preemptive action against potential attack vectors and better protect themselves. Organizations should ensure their defense systems are up-to-date with the latest threats and trends, as well as tailored to their specific needs. By taking these steps, they can more effectively combat malicious actors without compromising their security or privacy.

Psychological and behavioral traits of attackers

Attackers in the cybersecurity landscape often display psychological and behavioral traits that security professionals need to understand to devise effective defense strategies. Attackers are driven by their curiosity, enthusiasm for intellectual stimulation, and determination to outsmart security defenses.

One of the most common traits of attackers in the cybersecurity landscape is risk-taking behavior. Attackers are driven to take risks in order to achieve their objectives, even if it means overcoming obstacles or breaking existing laws or regulations. They are willing to push the boundaries and take risks because they believe that the rewards associated with success outweigh the potential consequences associated with failure. Another trait commonly seen among attackers is high levels of technical proficiency and knowledge. Attackers often have extensive experience with technology and software programming languages, which allows them to develop custom tools and scripts used for malicious purposes. Additionally, attackers typically possess detailed knowledge regarding networking protocols, cryptography algorithms, and operating systems; this allows them to understand how systems work so they can exploit weaknesses and bypass security measures.

Finally, attackers often demonstrate tenacity and persistence when attempting to compromise a system or network. Many times, attackers will continue trying different methods even after initial attempts have failed until they find a successful way into a target system. This type of behavior reflects an unwillingness on the part of attackers to give up easily on their objectives; instead, they persist until they succeed in achieving their goal. As such, they exhibit high creativity, innovation, and adaptability in devising new attack techniques and exploiting vulnerabilities. Attackers frequently employ deception and manipulation skills through social engineering tactics that exploit human weaknesses. Unfortunately, most attackers lack empathy or ethical considerations when pursuing their objectives.

Comprehending these psychological and behavioral traits is vital for organizations to anticipate potential attack vectors and create robust security controls. It allows organizations to design targeted countermeasures that keep ahead of malicious actors in the ever-evolving cybersecurity domain. Understanding these traits also serves as a reminder of the importance of continuously innovating one's defensive strategies while keeping in mind the mindset of attackers. By recognizing attackers' psychological and behavioral traits, organizations can develop appropriate mitigation measures for protecting their data and systems from malicious actors. Furthermore, this understanding can assist security professionals in preventing potential attacks before they occur by leveraging their knowledge of the mindset of attackers.

The role of the attacker's mindset in strengthening cybersecurity

Cybersecurity professionals must adopt an attacker's mindset to stay one step ahead of emerging threats. By understanding the tactics and strategies of adversaries, security teams can proactively identify and protect high-value assets, anticipate attack vectors, uncover hidden vulnerabilities, and conduct realistic simulations to test security controls. This involves combining traditional security solutions such as **endpoint platform protection (EPP)** and firewalls with proactive measures such as **identity threat detection response (ITDR)**, **cloud workload protection (CWP)**, and **cloud security posture management (CSPM)**. Security teams should also use analytics to detect anomalies in data and investigate suspicious activities, as well as implement identity management tools to restrict access privileges to authorized personnel only. Security teams should also implement encryption technologies to securely store and transmit sensitive data across networks. Finally, security teams must perform regular vulnerability scans to uncover hidden risks and send timely alerts whenever new threats emerge.

Furthermore, adopting an attacker's mindset encourages a culture of continuous improvement within organizations. Security teams must constantly challenge assumptions and explore new techniques to strengthen their defenses as cyber threats evolve rapidly. Security teams should conduct regular threat simulations to challenge assumptions and continually test their security controls. Attack simulations help organizations identify their weaknesses and improve their response capabilities. Additionally, performing frequent penetration tests can uncover hidden vulnerabilities in the system. Security teams should also regularly review their security policies to ensure they are up-to-date with industry standards and best practices. Finally, leveraging threat intelligence solutions can help organizations deter cybercriminals by proactively detecting malicious activities on networks before an attack occurs. By taking a proactive approach to security and continuously challenging assumptions, organizations can protect themselves against potential threats. This approach promotes exploring emerging technologies, techniques, and best practices so organizations can remain resilient against malicious actors.

To gain deeper insights into the attacker mindset, security professionals should look beyond traditional defense strategies and view potential risks differently. By thinking like an adversary, organizations can better assess which systems or data are most valuable to target, and prioritize resources accordingly. To feel like an adversary, security professionals need to examine the motivations and techniques of malicious actors. They can start by studying the recent attack trends, analyzing their strategies, and understanding how they gain access to networks or data. Security teams should also stay abreast of any changes in the threat landscape by leveraging resources such as blogs, conferences, or cybersecurity-related industry forums.

Additionally, they should monitor potential sources of attack indicators, such as social media sites or online discussion boards where attackers share their experiences and tactics. Finally, participating in **Capture the Flag (CTF)** events will allow security teams to practice their skills and expose them to various attack methods to help them anticipate attackers' behavior more effectively. By doing so, and combining this knowledge with existing defense strategies and technologies available today, security teams will be better prepared when defending against advanced attacks.

Organizations can further enhance their cybersecurity maturity by deploying specialized teams: the Blue Team focuses on defense, monitoring for threats, and safeguarding assets, while the Red Team simulates cyberattacks to identify vulnerabilities. The Purple Team, merging the roles of Red and Blue, uses both teams' insights to refine security strategies. This structure allows organizations to identify weaknesses, fortify defenses, and better prepare for cyber threats through focused, collaborative efforts.

Cybersecurity professionals must understand the adversary's mode of operation to identify the critical assets that are at risk and devise effective defense strategies against potential attacks. Adopting an attacker's mindset enables organizations to stay ahead of evolving threats by improving threat awareness, uncovering hidden vulnerabilities, conducting more realistic tests, and fostering innovation in defensive strategies.

Ethical considerations and legal boundaries

Ethical considerations and adherence to legal boundaries are of utmost importance. Security professionals must act within the confines of laws, regulations, and ethical standards to ensure their actions align with the rules governing accepted behavior. Integrity, respect for privacy, and protection of user rights are all critical components in ensuring ethical behavior. Responsible disclosure is a vital part of being an ethical cybersecurity professional – when security vulnerabilities are discovered, they should be reported to the affected parties or appropriate authorities rather than exploiting them maliciously or disclosing them publicly without proper coordination.

To maintain trust in their profession and avoid potential legal repercussions, security professionals must also be aware of applicable laws, regulations, and industry-specific requirements to remain compliant. This includes understanding data protection laws, intellectual property rights, and hacking, privacy, and incident response laws. Moreover, respecting privacy rights is essential; professionals should ensure proper safeguards are in place to protect the privacy and confidentiality of individuals and organizations by limiting access to those with authorized privileges and using it only for legitimate purposes. One example of applicable laws, regulations, and industry-specific requirements is the **Health Insurance Portability and Accountability Act (HIPAA)**. This act was established to protect the privacy and security of health data stored by healthcare providers, insurers, and other organizations.

Furthermore, ethical considerations relate strongly to hacking techniques; it is essential for security professionals not to engage in activities that could harm or violate privacy, even if their purpose is testing or improving security measures. When professionals prioritize adhering to ethical standards over other objectives, they build trust among stakeholders while contributing positively to cybersecurity community developments. Additionally, compliance with legal boundaries as well as ethical guidelines strengthens an organization's standing by providing credibility, which helps protect systems, data, and user privacy more effectively.

Ethical hacking and responsible disclosure

Ethical hacking is assessing the security measures of a system, network, or application to find and address any weaknesses or vulnerabilities to improve security. Ethical hackers use tools and techniques such as port scanning, vulnerability scans, password cracking, and social engineering to assess an organization's security. This form of testing helps organizations identify and address potential threats before they can be exploited by malicious actors. One benefit of ethical hacking is that it allows organizations to identify their vulnerabilities and take proactive steps toward addressing them. An additional benefit is that by hiring ethical hackers for their services, organizations can access the expertise they may not have internally. Finally, ethical hacking can help protect an organization's brand reputation by ensuring its customers' data remains secure.

On the other hand, responsible disclosure is another crucial aspect of ethical hacking involving notifying an organization's IT department when a vulnerability has been detected and allowing them time to patch it before disclosing it publicly. Responsible disclosure keeps potential attackers from exploiting these vulnerabilities while giving organizations time to fix the issue before it becomes public knowledge. Responsible disclosure requires these findings to be reported promptly to those affected or to the **computer security incident response team** (**CSIRT**). This approach works proactively, allowing organizations to recognize potential threats before malicious actors can exploit security flaws. Responsible disclosure is essential for maintaining a balance between security researchers, organizations, and users. It facilitates collaboration among researchers and organizations to enhance overall defense mechanisms. Organizations can further protect their reputation and the trust of their user base by swiftly and efficiently addressing found vulnerabilities with proper public notice.

Ethical hackers must adhere to responsible disclosure guidelines when presenting their findings, which include verifying the impact of identified vulnerabilities, documenting all data collected, and sharing information with applicable people inside the organization or through collaborative platforms. Following responsible disclosure guidelines allows organizations to act quickly on vulnerabilities before malicious individuals can hijack them. This decreases the risk posed by potential security incidents while strengthening cybersecurity defenses. Responsible disclosure also promotes a positive relationship between ethical hackers and organizations, inviting organizations to implement vulnerability disclosure programs where researchers can report findings without fear of legal repercussions.

In conclusion, ethical hacking and responsible disclosure are essential for keeping networks secure. They allow organizations to identify vulnerabilities proactively before malicious actors can exploit them, and provide a platform for collaboration between researchers and companies that ultimately results in stronger cybersecurity defenses. By following the guidelines of responsible disclosure when presenting findings, both parties can benefit from increased data protection while building relationships based on trust. With these measures in place, we can continue to ensure our online safety is not compromised.

The role of innovation in cybersecurity

Innovation plays an integral role in cybersecurity and is essential for organizations to address the ever-changing threat landscape. Always remember: as new technology becomes available, it is not just available to defenders but also to adversaries. Adopting innovative technologies and solutions allows organizations to stay ahead of cyber threats, improve their detection and prevention capabilities, and enhance their incident response and recovery systems. Emerging technologies such as AI, ML, quantum computing, big data analytics, blockchain, cloud-native security solutions, and IoT security advancements are being utilized to enable adaptive defense strategies.

However, some challenges must be addressed when implementing these innovative security solutions. Organizations need to consider privacy concerns, regulatory compliance issues, the impact on legacy systems, and a potential skill gap when innovating in this space. Therefore, organizations must foster a culture of continuous improvement while balancing security and usability.

Innovation in cybersecurity plays a vital role in protecting sensitive information and safeguarding digital assets from malicious attacks. By embracing creativity and team collaboration, organizations can proactively secure their networks against cyber breaches and remain resilient in the face of evolving threats.

The benefits of and need for innovation

Innovation is crucial in cybersecurity, offering numerous advantages and addressing the ever-evolving threat landscape. By incorporating innovative approaches, organizations can remain one step ahead of attackers and protect their assets, data, and reputation.

The constantly changing threat landscape requires creative solutions to detect, prevent, and respond to cyber threats. Organizations that embrace innovation can build cutting-edge solutions with the latest technologies to address new risks. Such technologies include AI, ML, and big data analytics. These technologies enhance cybersecurity capabilities by improving the accuracy of threat detection and analysis.

Security professionals should assess and implement various innovations to keep up with the ever-evolving threat landscape. The top three essential innovations they should evaluate and implement include AI, ML, and big data analytics. AI provides organizations with a more comprehensive view of potential threats by continuously learning from the environment and adapting to it. ML can detect threats faster by leveraging previous data points and predicting future attacks before they happen. Lastly, big data analytics allows for efficient data analysis, which increases accuracy in identifying vulnerabilities or malicious activities before it's too late. By incorporating these innovative approaches, security professionals can remain one step ahead of attackers while protecting their businesses from cyber risks.

Additionally, innovation encourages **security by design**. This proactive approach involves integrating security measures during the early stages of system and software development, thus reducing vulnerabilities and strengthening the overall cybersecurity posture. Moreover, automation can be implemented through innovation – automating tedious security processes while speeding up incident response time, thus reducing manual errors or mistakes. AI-driven solutions also provide businesses with a more comprehensive view of potential threats as they can continuously learn and adapt, leading to higher accuracy in identifying potential vulnerabilities or attacks before they occur.

Innovation is essential for organizations looking to remain secure against evolving cyber threats. By leveraging emerging technologies through innovative approaches such as automation or security by design principles, businesses can anticipate potential threats more efficiently while minimizing their attack surface area, ultimately protecting their assets, data, and reputation from malicious actors.

Driving innovation within organizations

Innovation plays a critical role in cybersecurity by providing the necessary tools and methodologies to counteract evolving cyber threats. As malicious actors continuously develop more sophisticated techniques, innovation ensures that defensive strategies and technologies keep pace. This includes the development of advanced algorithms for threat detection, the integration of ML for predictive analytics, and the creation of more robust encryption methods. Additionally, innovation drives the creation of user-friendly security solutions, making it easier for individuals and organizations to implement strong cybersecurity measures. By constantly pushing the boundaries of what is possible in cybersecurity technology and practices, innovation acts as a crucial defense mechanism against an ever-changing array of cyber risks.

The healthcare sector has adopted advanced encryption methods, secure data storage systems, proactive monitoring tools, telemedicine, and remote patient monitoring technologies to ensure the privacy and integrity of confidential information. In the financial sector, banks and other financial institutions have implemented biometric authentication solutions, such as fingerprint and facial recognition, to strengthen their customer authentication processes and mitigate the risks of identity theft and fraud. In the technology industry, software development companies have incorporated certain coding practices into their innovation processes. By adhering to industry best practices, employing code review techniques, and utilizing automated vulnerability scanning tools during development, organizations can identify existing security flaws before their products hit the market. Additionally, cloud-native security solutions are being deployed in various industries to protect data and infrastructure stored in the cloud. These security measures include **cloud access security brokers** (**CASBs**), container security, and serverless security, significantly reducing potential breaches while providing businesses with greater control over sensitive information.

Innovation in cybersecurity is vital for organizations to remain secure against emerging threats. Through implementing innovative technologies such as biometric authentication, advanced encryption methods, secure coding practices, and cloud-native security solutions, organizations can benefit from improved customer authentication processes while protecting sensitive data from sophisticated cybercrime tactics.

Emerging technologies and future trends

Cybersecurity is constantly evolving due to the emergence of new technologies and trends. Examples of such innovations include generative AI, big data analytics, quantum computing, and blockchain. Always remember that as emerging technologies arise, threat actors similarly are evaluating them for their use against defenders. These technologies have revolutionized how organizations handle cybersecurity threats by providing better visibility and control over network activity, data encryption protocols, anomaly detection algorithms, digital identity systems, and post-quantum cryptography.

Generative AI, encompassing **large language models** (**LLMs**), is capable of generating content and responses from extensive training data. Vendors such as SentinelOne are now offering solutions that utilize this technology, enabling virtual SOC analyst capabilities and guided hunting through Purple AI.

- **Quantum computing-based security**: Quantum computing-based security solutions could offer a range of advantages for security professionals. Using quantum mechanics' power, these solutions can process data faster than traditional computers. This can allow organizations to identify threats faster and respond more quickly.

- **Big data analytics**: Big data analytics can also be used to understand user behavior better, helping organizations detect suspicious activity and insider threats. As a result, big data analytics enables security professionals to make more informed decisions about promptly responding to potential threats. Lastly, big data analytics can provide new insights into cyber-attacks as they happen, allowing organizations to understand better what type of attack is taking place and how best to mitigate the threat.

- **Blockchain-powered security**: The distributed ledger system built into blockchain helps protect transactional data by providing an immutable record of activity that can't be altered or manipulated. Cryptography is used to secure this data at each step of the process, which allows organizations to enjoy further protection against malicious actors. Blockchain-based systems are inherently resistant to various cyber-attacks, such as DDoS and man-in-the-middle attacks.

By staying up-to-date with these emerging technologies and trends in cybersecurity, by implementing tools such as generative AI algorithms, IoT security protocols, cloud-native solutions, blockchain technology platforms, post-quantum cryptography methods, and user behavior monitoring systems, businesses can defend themselves from malicious actors more effectively. Ultimately, these advancements and innovations enable organizations to create robust cybersecurity defense strategies that keep them one step ahead of attackers.

Summary

We conclude this chapter by highlighting the need for organizations to take an innovative approach in order to stay ahead in the constantly evolving cybersecurity landscape. It emphasizes the importance of remaining constantly vigilant and well-informed, similar to operating a radar system, in order to effectively anticipate and mitigate cyber risks. The chapter underscores the advancements made by adversaries when it comes to attack methodologies, as they often adopt technologies more rapidly than defenders can keep up with. This situation necessitates an approach where understanding types of cyber threats, including those that leverage advanced technologies such as AI and cloud computing, becomes imperative. The chapter also emphasizes the significance of cybersecurity strategies, such as employing threat-hunting methods to swiftly and effectively detect and respond to potential threats.

Moreover, this chapter advocates for a forward-thinking cybersecurity posture that goes beyond keeping up with technological advancements. It calls for organizations to prioritize training and preparedness while emphasizing innovation and adaptation in their cybersecurity practices. The chapter serves as a reminder of the sophisticated nature of cyber threats, highlighting the essential role of strategic foresight and continuous learning in maintaining strong defenses against these challenges. By combining vigilance, innovation, and a comprehensive strategy that encompasses prevention and preparedness measures, organizations can better protect their digital assets against the complexities of the cybersecurity domain.

4

Assessing Your Organization's Security Posture

This chapter is tailored to provide insights for assessing your organization's overall security posture, considering three critical facets: technology, processes, and people. The chapter delves into the essentials of a comprehensive security posture, illuminating the multifarious components that constitute robust security, including hardware, software, and the human factor. It further explores the metrics that effectively measure the performance and effectiveness of security programs and teams, providing a quantifiable perspective on cybersecurity. Lastly, the chapter underscores the significance of asset inventory management in solidifying your security posture and discusses the strategies for the continuous monitoring and enhancement of the security landscape.

This chapter will cover the following topics:

- The components of a comprehensive security posture
- Effective metrics for security programs and teams
- Asset inventory management and its role in security posture
- Continuously monitoring and improving security posture

Let us dive in!

The components of a comprehensive security posture

Everyone talks about the balance between people, processes, and technology, but the question is, how do we achieve it? Let's dive into learning the key components of a comprehensive security posture. We begin by exploring the array of technological tools and defenses contributing to a robust cybersecurity landscape. In doing so, we will not only identify these elements but also evaluate their effectiveness and alignment with the organization's unique security needs. Subsequently, we will switch gears to examine the procedural aspect of cybersecurity, highlighting the significance of establishing and consistently implementing well-documented security processes. Lastly, we will focus on the most unpredictable yet crucial component—people. Recognizing the substantial impact of human behavior on security, we will stress the importance of cultivating a security-aware culture through effective training programs. Furthermore, we will navigate the potential risks stemming from human factors and discuss effective strategies to mitigate these risks.

Evaluating security technologies

When evaluating security technologies, it is essential not to fall for marketing. Every vendor out there claims to increase an organization's security posture and improve any possible metric you might have inside your security department. While for some that might be accurate, you shouldn't take the vendor's claim for granted and should instead always do your own assessment.

Effectively gauging the efficacy and alignment of your organization's security technologies is imperative to managing your security posture. The presence of sophisticated tools alone does not ensure security; they must be continuously evaluated and tailored to fit your organization's specific needs. This process ensures the relevancy and timeliness of your technological defenses and optimizes their efficiency.

To start, it is crucial to define your organization's security requirements. Your organization should define these requirements, not the vendor trying to make the sale. When defining security requirements, there are many things to consider, including regulatory and compliance requirements. For example, a financial institution that handles sensitive customer data may have different security requirements from a retail company that mainly operates inventory data.

Additionally, factors such as your organization's size, cybersecurity maturity, available skills, industry regulations, and the nature of the data you manage must be considered. This deep understanding must form the backbone of your evaluation process and set the standard for your effective security strategy.

Once we understand the security requirements for our organization, we can dive into the deployment and configuration of security technologies. Always remember that the default configuration of your technology might not be the most secure. For example, validate that your firewalls are configured to block all unwanted incoming traffic or that there are no unnecessary ports open that the adversary could exploit.

Similarly, validate that you have the means of effectively patching the operating system and applications in your environment when new security patches are available. Also, consider whether your XDR solution is ingesting all your security data to detect and respond across your digital estate automatically. The accurate deployment and configuration of security tools warrant scrutiny and should never be rushed.

Always remember that cybersecurity is a rapidly evolving field. There will always be more attacks this month compared to the previous, and there will always be more connected devices this year than last year. Keeping up to date with advancements in cybersecurity technologies and implementing them timely is crucial. Consider the integration and interoperability of your security tools as well. For instance, your SIEM system should seamlessly collate and analyze data from your firewalls, IDS/IPS, and other security tools. An integrated security infrastructure enhances the visibility of your security landscape, making it easier to detect, respond to, and mitigate threats.

Lastly, your evaluation should encompass a review of your incident response capabilities. For example, if a cyber-attack is detected, does your security technology stack enable a rapid and coordinated response? Is there an automated containment procedure in place to prevent the spread of the intrusion? The ability to quickly detect, respond to, and recover from security incidents is crucial to your security technologies' effectiveness.

By conducting this comprehensive evaluation, you can pinpoint gaps, redundancies, and opportunities for improvement in your security technologies. This proactive approach is vital to maintaining a robust defense mechanism to thwart evolving cyber threats.

Understanding the role of security processes

Processes are critical to an effective cybersecurity strategy and architecture. The importance of processes should never be downplayed. For example, consider an incident response plan as you would your organization's fire drill procedure. The purpose of fire drills is to prepare people to evacuate a burning building safely. A well-prepared incident response plan, on the other hand, guides an organization to rapidly detect, contain, and remediate a security breach. Without it, the organization may be thrown into chaos during a breach, akin to people running aimlessly during a fire, leading to greater damage. Thus, the incident response process is critical in ensuring an organized, effective response to minimize the impact of cybersecurity incidents. Well-thought-through and well-formulated security policies set the stage for an effective execution. There are several widely recognized references for developing effective security policies. The **National Institute of Standards and Technology** (**NIST**) provides comprehensive guidelines, including the NIST Special Publication 800-53, which covers security and privacy controls. The **International Organization for Standardization** (**ISO**) and the **International Electrotechnical Commission** (**IEC**) have compiled ISO/IEC 27002, an international standard offering best practices for information security management. In addition, the **Information Systems Audit and Control Association** (**ISACA**) provides the **Control Objectives for Information and Related Technology** (**COBIT**), a framework for IT governance and management. Lastly, the **Center for Internet Security** (**CIS**) offers the CIS Controls, a prioritized set of actions that collectively form a defense-in-depth set of best practices for cybersecurity. Each of these references provides invaluable guidance for creating robust security policies and should be taken into consideration when formulating your organization's security strategy. These policies must clearly outline the purpose and provide the clarity required on the roles and responsibilities of employees, what has to happen in which situation, and how.

For instance, a password policy might dictate that passwords should be a mix of letters, numbers, and special characters, changed every 60 days, and not repeated within a year. On the other hand, a **Bring Your Own Device** (**BYOD**) policy might specify that personal devices accessing company data should have approved antivirus software installed and that data should not be stored on personal devices without encryption.

Risk assessments enable organizations to perform periodic exercises that proactively identify potential vulnerabilities and assess their impact. For instance, a risk assessment might reveal a company's customer database is at risk due to an outdated database management system. The company could then prioritize updating the system or implementing additional safeguards.

Incident response planning is another key process, laying out a detailed action plan for dealing with security incidents. For example, a plan might specify that upon detecting a potential data breach, the incident should immediately be reported to the IT department, the suspected system should be isolated from the network, and the nature and extent of the breach should be documented for further investigation and resolution.

Regular audits complete the set of core security processes. These are independent assessments that measure the effectiveness of security controls and identify gaps. For instance, an audit might reveal that an organization's backup processes are inadequate, with backups taken infrequently and only stored on-site, raising concerns about data loss in the event of a physical disaster at the site.

Overall, security processes provide the guiding framework for all cybersecurity activities in an organization. They establish the rules, guide risk management, spell out incident response procedures, and, through audits, ensure a cycle of continuous review and improvement. With robust and consistently enforced security processes, an organization can safeguard its cyber environment effectively.

The human factor in a security posture

Humans are often described as the weakest link in an organization's security strategy. This is because the human factor is often an organization's most unpredictable and potentially vulnerable aspect. Regardless of whether we talk about employees, third-party partners, or contractors, they all can expose a significant source of risk to an organization, intentionally or unintentionally.

This is why employee awareness and training are critical components of your security posture and shouldn't be considered optional or something you just have to do. Employees must become aware of the cybersecurity threat landscape and understand the possible impact of clicking on a malicious link or opening a malicious file while being encouraged to report suspicious activities without the fear of being retaliated against because they might have fallen for a sophisticated social engineering technique. Regular training programs can help to educate them about the latest threats and how to respond appropriately.

In addition to employee awareness and training, clear communication of security policies is vital. Security policies shouldn't be buried under sub-sections of a central repository that the employees vaguely know of existence; instead, they should be widely accessible, easy to understand, and regularly updated to reflect changes to the threat landscape and the organization's security posture. Every employee should know their role in maintaining the organization's security.

Access controls also play a vital role. Ensuring that employees have access only to the information and systems they need for their work can minimize the risk of insider threats and reduce the potential damage if an employee's account is compromised. While considering access controls, it's important to factor in conditional access, ensuring that only trusted identities have access to corporate resources and services when needed instead of access by default.

Lastly, an open and transparent security culture encourages employees to report potential security issues without fear of blame. An effective security culture can turn your employees from potential security risks into a line of defense against threats.

In conclusion, people are essential to an organization's security posture. Organizations can mitigate the risks associated with the human factor by fostering a security-conscious culture, maintaining robust access controls, and providing regular training and clear communication of policies. In the following section, we aim to demystify the complex realm of security metrics, guiding you through their selection, implementation, and analysis, so you can utilize them as potent instruments for strengthening your cybersecurity defenses.

Effective metrics for security programs and teams

Metrics serve as tangible indicators of the efficacy of your security programs and are a vital part of maintaining, improving, and communicating your security posture. But the question is which metrics are required, how can we effectively report on these metrics, and which metrics actually help improve the security posture instead of just looking nice on a business analytics dashboard?

The right metrics can provide insights into trends, reveal vulnerabilities, and enable you to make data-driven decisions to enhance your cybersecurity defenses. However, not all metrics are created equal. Choosing the right metrics that accurately represent the state of security and align with your business objectives is an art in itself. Once chosen, they must be consistently tracked, analyzed, and used as a basis for continuous improvement.

In this section, we will delve into the importance of security metrics, the process of selecting meaningful metrics, the implementation and tracking of these metrics, and finally, how to analyze them for constant improvement. Our goal is to equip you with the knowledge to transform these abstract numbers into powerful tools in your cybersecurity arsenal.

Understanding the importance of security metrics

The purpose of security metrics is to provide quantifiable evidence of the effectiveness of your organization's security posture. They serve as the bridge between complex cybersecurity operations and comprehensible business insights that can guide decision-making processes. Security metrics give a pulse on the health of your cybersecurity defenses.

Consider, for example, a metric such as *number of unpatched vulnerabilities*. By tracking this metric over time, you can understand the speed and effectiveness of your patch management processes. An increasing trend might signify a need for more resources or a review of the existing patch management process. In contrast, a decreasing trend can validate the effectiveness of your current efforts. The metric *average time to detect a security incident* in an incident response scenario can provide valuable insights. If this time is too long, it indicates a gap in your detection capabilities, potentially requiring improvements in your monitoring systems or anomaly detection algorithms.

Additionally, security metrics can help align cybersecurity efforts with business objectives, enhancing communication with stakeholders. For instance, translating *potential cost of a data breach* into projected financial losses or reputational damage can convey the importance of investing in cybersecurity to senior management more effectively.

Moreover, regulatory and compliance requirements often necessitate the use of security metrics. For example, to comply with GDPR, organizations must demonstrate that personal data is protected, and metrics such as *number of data breaches* and *average time to respond to a data breach* can provide tangible evidence of compliance.

In essence, security metrics, when properly defined, tracked, and analyzed, can drive the strategic direction of your cybersecurity efforts. They provide a clear, quantifiable way to evaluate the effectiveness of your security controls, align cybersecurity goals with business objectives, and comply with regulatory requirements. By translating the complex world of cybersecurity into understandable and actionable insights, metrics empower you to manage your organization's security posture proactively and effectively.

Selecting the right metrics

Metrics that are overly complex or lack context can be misleading and lead to misinterpretation. This is why good metrics must be easy to understand and communicate. For example, a metric such as *percentage of employees who passed the annual cybersecurity awareness test* is easy to understand and provides a clear insight into the effectiveness of your cybersecurity training program.

Effective security metrics should be aligned with your organization's business goals and security objectives. For instance, if your organization prioritizes customer data protection, metrics such as *number of data breaches involving customer data*, *average time to detect data breaches*, and *average time to respond to data breaches* might be particularly pertinent.

Additionally, you should also consider the scope of the metrics. Operational metrics such as *number of detected malware incidents* offer insights into day-to-day security operations. Strategic metrics such as *potential financial impact of identified risks* provide a higher-level view of the organization's security posture and can be useful in discussions with senior management.

One approach for selecting the right metrics is to use frameworks such as the NIST Cybersecurity Framework or the CIS Controls. NIST and CIS are two widely respected models in the cybersecurity field. Both offer detailed and structured approaches to cybersecurity and can be immensely useful in guiding your selection of security metrics:

- **CIS Controls:** The CIS Controls are a series of recommended actions that offer a robust defense against the prevalent forms of cyber attacks on systems and networks. For each control, you can define specific metrics. For instance, the first control is Inventory and Control of Hardware Assets. A possible metric for this control could be *percentage of unauthorized devices detected on the network*. This quantifiable metric aligns directly with the objective of the control and clearly indicates how well you are performing against it. *Figure 4.1* visualizes the CIS Controls.

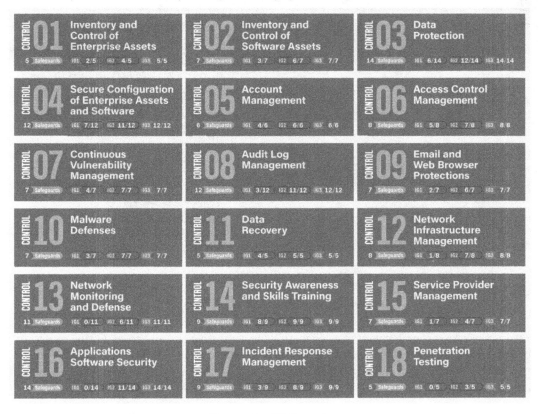

Figure 4.1 – CIS Controls

- **NIST Cybersecurity Framework**: The NIST framework consists of standards, guidelines, and best practices to manage cybersecurity risk. It is divided into five core functions: Identify, Protect, Detect, Respond, and Recover. Each function includes several categories and subcategories. For example, under the Detect function, there is an Anomalies and Events category. A relevant metric for this category might be *average time to detect a security anomaly*. This metric clearly indicates your detection capabilities and aligns directly with the aim of the Detect function. *Figure 4.2* demonstrates the NIST Cybersecurity Framework across the 5 stages.

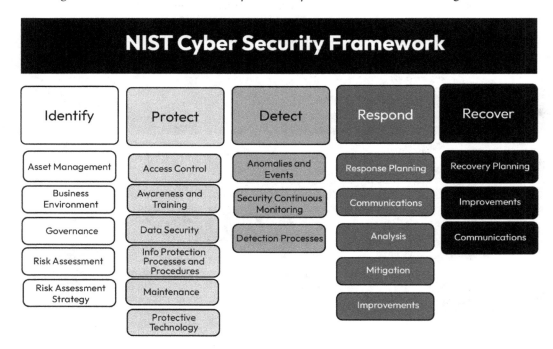

Figure 4.2 – NIST Cybersecurity Framework

These frameworks ensure that your selected metrics are comprehensive and cover all vital aspects of your cybersecurity posture. They help to ensure that your metrics are not randomly chosen but are systematically tied to respected industry standards, thus making your metric program more robust and reliable.

Remember, no single metric can provide a complete picture of your organization's security posture. Therefore, it's essential to select a balanced set of metrics that collectively provide a comprehensive view of your security performance. By choosing metrics that are aligned with business goals, easy to understand, appropriately scoped, and that encourage positive behavior, you'll be well equipped to effectively measure and manage your organization's cybersecurity.

Implementing and tracking security metrics

Implementing and tracking security metrics is an integral part of ensuring the safety and security of an organization. To do this effectively, it is necessary to identify data sources that will provide insight into the desired metric, set up automated processes for collecting this data to ensure accuracy and consistency, record it at regular intervals, and present it in an easily interpretable format.

Data sources include firewalls, intrusion detection systems, and organizational databases that record user behavior or system logs. Depending on what metric is being tracked, it may be necessary to combine multiple sources to get a comprehensive view. For instance, measuring *time to patch vulnerabilities* would require data from a vulnerability management tool, patch management system, and incident response system.

To automate collecting data for these metrics, SIEM systems or automated reporting features built into security tools can be employed. Custom scripts or programs may also be used to pull information from different sources. The frequency with which the data should be recorded will depend on the tracked metric type and organizational needs; some may require daily collection, while others might only need quarterly recordings.

Once collected, the data should be presented to make it easy for stakeholders to understand and interpret trends over time while also allowing them to drill down into specific areas if needed. Dashboards are often employed for this purpose since they offer visual representations of the metrics and interactive elements so users can quickly interact with them. However, simply implementing and tracking metrics is not enough; regular reviews should also be performed to ensure their relevance in assessing an organization's security posture given any changes in its environment or threat landscape.

Asset inventory management and its role in security posture

As we embark on the third section of our exploration into assessing an organization's security posture, our attention turns to the critical role of asset inventory management. The bedrock of any successful cybersecurity strategy is a comprehensive understanding of what you are protecting – your assets. These assets can range from physical hardware and digital systems to data and human resources. In this section, we will delve into the importance of asset inventory in cybersecurity, the process of building a comprehensive asset inventory, strategies for maintaining and updating this inventory, and finally, the role of asset inventory in effective risk management.

Understanding asset inventory in cybersecurity

You can't defend what you don't know you have. As such, you must have as much visibility as possible into your assets regardless if we talk physical such as workstations, laptops, digital assets such as software, data, intellectual property, or human assets, who are the people in your organization who access your corporate resources and data. Your assets are what you are trying to protect from potential threats, be it cyber criminals, insider threats, or inadvertent user errors.

Suppose you are a security officer at a healthcare organization. Your physical assets would include the computers and servers where patient records are stored, the network equipment that facilitates data transfer, and the mobile devices used by staff. Your digital assets include the **Electronic Health Record (EHR)** software, patient data, and proprietary algorithms for predicting patient outcomes. Your human assets include your medical staff, who access patient records; your IT team, who manage the servers and network; and any third-party vendors with access to your systems. Understanding your assets is crucial in many aspects of cybersecurity. It enables you to identify what needs to be protected, where your vulnerabilities might lie, and the impact of a security incident.

For instance, knowing that your servers are running an outdated operating system can alert you to potential vulnerabilities that need patching. Understanding that your patient data is stored on these servers, and the legal and reputational implications of a data breach, can help you assess the risk and prioritize this patching. Recognizing that your medical staff often access patient records from their personal devices can highlight a potential vector for data leakage or malware infection.

A comprehensive asset inventory forms the basis of your cybersecurity efforts. Providing a clear view of your organization's landscape allows you to build a robust, targeted, and effective security strategy.

Building a comprehensive asset inventory

Creating a comprehensive asset inventory is an essential first step toward a solid cybersecurity posture. This process involves identifying and cataloging all assets within an organization, accompanied by relevant information about each asset. Historically, this had been a daunting task, but a systematic approach can make the process manageable and effective.

The first step for an effective mean of asset inventory is, of course, identification. This involves a detailed sweep of the organization to list all assets. For physical assets, this might involve physical checks, consulting purchase records, and speaking with department heads. Network scanning tools, software inventories, and database schemas could be useful for digital assets. HR records, contractor lists, and departmental rosters can provide the necessary information for human assets.

For example, a retail corporation may have hundreds of **Point of Sale (POS)** systems (physical assets) across different stores. Each system runs POS software (digital assets) and is operated by store employees (human assets). A thorough identification process will uncover all these assets. Once identified, each asset needs to be categorized. Categories might include hardware, software, data, and people and can be further divided into subcategories. For example, hardware can be broken down into servers, laptops, mobile devices, and network equipment. Subsequently, for each asset, pertinent information should be recorded. This could include location, owner, users, configuration, and the data it holds for physical and digital assets. For human assets, it might involve their role, level of access, and security training status. Returning to our retail corporation example, a POS system in a specific store will be cataloged with its location, the vendor details, the software version it runs, and the employee responsible for its operation. Detailed information such as this contributes to a more comprehensive and effective asset inventory.

Many automated discovery and inventory management tools are available to assist with the asset inventory process. Fortunately, there is a technology that can help you provide real-time insights into your asset library. Technology such as **Unified Endpoint Management (UEM)**, **Extended Detection Response (EDR)**, **TVM**, or **Cloud Security Posture Management (CSPM)** systems can help. These tools can automatically identify and catalog digital assets, track changes, and often offer visualization tools to help understand the asset landscape. However, these should be used in conjunction with manual checks to ensure completeness and accuracy. By building a comprehensive asset inventory, you equip your organization with a fundamental understanding of what is at risk, enabling effective vulnerability management, incident response, and security policy development. This thorough knowledge forms the backbone of your cybersecurity defense strategy.

Maintaining and updating asset inventory

The dynamic nature of organizations today, with new assets being added and old ones being retired regularly, necessitates the continuous maintenance and updating of the inventory. An outdated or inaccurate inventory can leave your organization vulnerable to overlooked security risks.

Routine audits are an effective strategy for maintaining your asset inventory. These audits involve a systematic review of your inventory to verify that it is complete and accurate. They can identify assets that have been missed in the initial sweep or have been added since. For example, an IT services company might conduct a quarterly audit to verify that all new servers or software licenses are accounted for and that retired assets are removed from the inventory.

Automated asset management tools can also be valuable in maintaining your inventory. These tools can often detect when new assets are added to your network, track changes to existing assets, and alert you to potential discrepancies between the inventory and the actual asset landscape.

For instance, a financial institution using automated tools would be alerted when a new system is added to the network or when an existing server is upgraded. This real-time update ensures the asset inventory remains accurate and current, contributing to a strong security posture.

It's also essential to have processes to update the inventory when assets enter or leave the organization. This might involve IT and HR teams working together to ensure that when an employee joins or leaves, or when a new server is installed or old equipment is retired, the asset inventory is promptly updated.

Consider a manufacturing firm that frequently upgrades its machinery (physical assets), updates its proprietary design software (digital assets), and experiences high staff turnover (human assets). The firm can ensure that its inventory remains accurate and its security risks are effectively managed by establishing procedures for updating the asset inventory whenever these changes occur.

Maintaining and updating your asset inventory is not a one-time effort but an ongoing process. By committing to this, your organization can ensure that its security strategies and defenses protect all its assets effectively, enhancing your overall security posture.

Continuously monitoring and improving your security posture

By taking a proactive approach to security, organizations can gain real-time insights into potential issues and quickly mitigate them. To achieve this, continuous monitoring must be employed to identify emerging threats. By embracing continuous monitoring, you gain real-time visibility into potential security threats, allowing for swift response and mitigation.

Implementing continuous monitoring practices

Continuous monitoring involves the real-time collection, analysis, and interpretation of security-related data to detect and respond to threats promptly. For example, implementing an EDR allows for the continuous monitoring of endpoint activity and the detection of suspicious activities or attempted intrusions.

Log analysis is another crucial component of continuous monitoring. By aggregating and analyzing logs from various systems, such as firewalls, servers, and applications, security professionals can identify anomalous patterns or indicators of compromise. This process can provide early warnings of potential security incidents. Threat intelligence feeds are invaluable in continuous monitoring. They provide up-to-date information on emerging threats, attack vectors, and malicious activities observed across the industry. Integrating threat intelligence feeds into your monitoring systems enables proactive identification and mitigation of potential risks. *Figure 4.3* shows a practical example in this case from Mandiant's Advantage Threat Intelligence solution.

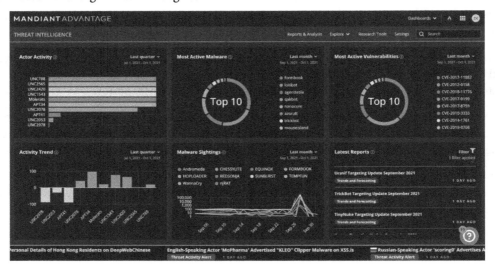

Figure 4.3 – Mandiant Advantage Threat Intelligence

Real-world examples highlight the importance of continuous monitoring. Consider the Equifax data breach, where unauthorized access went undetected for months due to inadequate monitoring. A robust continuous monitoring program could have detected unauthorized activity and allowed an immediate response, minimizing the impact of the breach.

Continuous monitoring also enables the identification of internal threats. For instance, an employee with access to sensitive data downloading large amounts of information may indicate potential data exfiltration. Real-time monitoring can flag such activities and trigger an investigation before significant damage occurs.

By implementing continuous monitoring practices, organizations can proactively detect and respond to security incidents, reduce incident response times, and minimize the potential impact of breaches. It provides security professionals with actionable insights and a proactive approach to maintaining a solid security posture.

Responding to incidents and implementing remediation measures

An incident response plan is essential for organizations to ensure they are prepared during a security incident. This plan provides a coordinated and efficient response to minimize the impact and facilitate swift remediation. It outlines the roles and responsibilities of key personnel, establishes communication channels, and defines steps to minimize the impact. *Figure 4.4* visualizes the process of information sharing throughout an incident lifecycle.

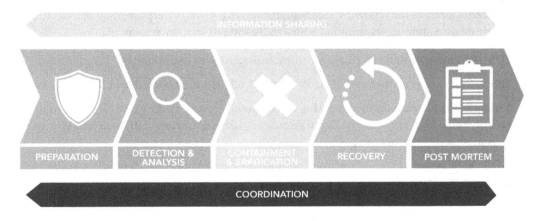

Figure 4.4 – Example workflow of incident response plan

When an attack occurs, **Incident Response Teams (IRTs)** are crucial in identifying, containing, and restoring affected systems. The SolarWinds supply-chain compromise attack is a recent example of how important it is for organizations to respond to incidents quickly.

Remediation measures must be implemented once an incident has been contained to address any vulnerabilities or weaknesses. This could involve patching systems, strengthening access controls, or revising security policies and procedures. In addition, post-incident analysis should be conducted to understand what caused the attack and identify areas for improvement that may lead to corrective actions, such as introducing multi-factor authentication protocols, delivering phishing awareness training programs, or increasing email filtering services.

Responding effectively to incidents and implementing remediation measures allows organizations to mitigate any potential damage from security incidents while improving their overall security posture. By learning from these incidents, organizations can adapt their defenses and increase their resilience against future threats.

The technological landscape in security posture

Technology forms the frontline defense of an organization's security posture. As such, it is vital to understand the various security tools and infrastructures that are critical for an organization. This section overviews these essential elements and their function in your security framework. Technologies range from traditional components such as **Intrusion Detection Systems (IDSs)**, **Intrusion Prevention Systems (IPSs)**, and firewalls to more advanced tools such as **Extended Detection Response (XDR)**, **Threat and Vulnerability Management (TVM)**, and **Security Information and Event Management (SIEM)** systems. Each of the following technologies serves a distinct purpose in an organization's security posture providing an organization with a multi-layered security system:

- **IDS**: IDSs monitor network traffic for suspicious activity and send alerts.

- **IPS**: IPSs go a step further than IDSs by automatically blocking potentially harmful activity.

- **Firewall**: The firewall controls incoming and outgoing network traffic based on predetermined security rules. They form the first line of defense in a network's security infrastructure, protecting it from unauthorized access.

- **XDR**: XDR systems extend detection and response capabilities across multiple surfaces such as endpoint, cloud, and identity.

- **TVM**: TVM systems help organizations understand their application inventory and the risks associated with the application's installed versions by comparing against known vulnerabilities and exploits while also giving remediation capabilities to patch insecure applications.

- **SIEM**: SIEM systems collect and aggregate log data generated throughout the organization's technology infrastructure, providing real-time analysis of security alerts generated by applications and network hardware.

While historically, security teams looked at technology in isolation and sometimes considered it a checklist, this model is becoming increasingly obsolete and replaced by the concept of the **cybersecurity mesh**, a relatively new addition to the technological landscape. It fundamentally redefines how we approach cybersecurity, enabling a more flexible, scalable, and reliable security architecture. Traditionally, security perimeters were built around the organization's network, often called the castle-and-moat approach. However, with the rise of remote work, IoT devices, and cloud computing, the perimeter has become increasingly fragmented, rendering the old approach ineffective.

The cybersecurity mesh addresses this by allowing the security perimeter to be defined around the identity of a person or thing. It enables each access point to be treated individually, enhancing the security of complex and geographically dispersed organizations. Cybersecurity mesh architectures deliver security controls where they are most needed, facilitating a more modular and responsive security environment. Understanding and incorporating this concept can significantly enhance the adaptability and effectiveness of an organization's security posture. In *Figure 4.5*, you can see an example of a cybersecurity mesh architecture.

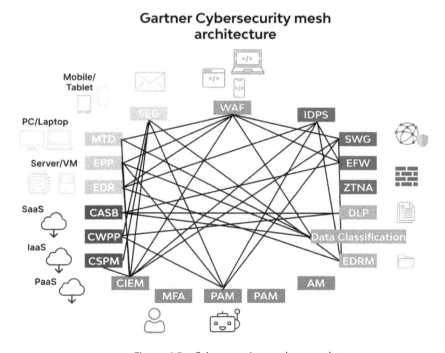

Figure 4.5 – Cybersecurity mesh example

Knowing that the effectiveness of your security tools and infrastructure investments depends on a profound and well-thought-through security architecture is essential. Understanding these technologies and their roles allows security professionals to construct and maintain a robust technological defense against potential cyber threats.

Summary

In the pursuit of a robust cybersecurity posture, the establishment and continuous maintenance of a comprehensive asset inventory is critical. This inventory, which catalogs all assets within an organization, forms the backbone of effective vulnerability management, incident response, and security policy development. It involves not only the identification and categorization of assets but also the recording of pertinent information about each one, with automated tools often playing a significant role in facilitating this process. However, to remain accurate and current, the inventory requires regular audits and real-time updates when assets enter or leave the organization.

Beyond asset management, it's crucial for organizations to proactively monitor security threats in real time and employ a solid incident response plan. Continuous monitoring, comprising practices such as endpoint activity surveillance and log analysis, can detect and respond to threats promptly, reducing incident response times and minimizing potential risks. In the face of a security incident, the incident response plan guides a coordinated and efficient response to minimize impact and facilitate swift remediation. This includes identifying and containing the incident, implementing remediation measures, and conducting post-incident analysis to adapt defenses and increase resilience against future threats. Thus, the combination of comprehensive asset management, continuous monitoring, and effective incident response forms the cornerstone of a strong cybersecurity defense strategy.

In the next chapter, you will learn the key elements of a successful cybersecurity strategy, enabling you to develop a comprehensive and modern cybersecurity strategy.

5
Developing a Comprehensive Modern Cybersecurity Strategy

Cybersecurity is a complex world, and it is not easy to find your way through.

This chapter will demonstrate how old strategies have become insufficient to battle on the levels of complexity. For instance, the traditional approach of creating a strong perimeter defense system, often referred to as a "moat and castle" strategy, was used to protect internal networks. This strategy, while effective against certain threats, fails to account for more sophisticated attacks and internal threats. Therefore, this chapter is not a refresher but rather shows a new perspective on how cybersecurity can be better intertwined with core business goals.

In this chapter, we will re-examine risk management, elevating it from a backseat role to a central strategic position. Next, we will dive into digital forensics and incident response, shedding light on their pivotal roles in today's cyber environment. Additionally, we will touch upon security awareness programs, highlighting the crucial role every team member plays in cybersecurity.

This chapter is designed to stimulate deeper discussions, foster valuable insights, and gather practical takeaways on the challenges and opportunities in our digital age.

This chapter will cover the following topics:

- Key elements of a successful cybersecurity strategy
- Aligning cybersecurity strategy with business objectives
- Risk management and cybersecurity strategy
- Incident response planning and preparedness
- Security awareness and training programs

Let's dive in!

Key elements of a successful cybersecurity strategy

A strong cybersecurity strategy is about understanding the complicated balance between risk, business objectives, and the constantly evolving threat landscape, rather than simply having a given defense mechanism or the latest technology in place.

As seasoned professionals would endorse, the strength of a strategy will not be found within individual components, but in how its components seamlessly integrate. When developing a cybersecurity strategy, we should consider questions such as, how do foundational principles align with the larger organizational ethos? What objectives should be set, and how will they be coherent with the organization's mission and vision? And, crucially, how do we determine the significance of each strategy element, ensuring that nothing is overlooked? With insights from real-world scenarios and forward-thinking approaches, we can embark on a journey to master the art and science of cybersecurity strategy.

Foundational principles and components

A robust foundation is crucial in the cybersecurity territory. Building a cybersecurity strategy is like building a skyscraper: without strong groundwork, even the most impressive structures can fall. When developing a comprehensive modern cybersecurity strategy, this foundation consists of the key principles and components.

Defense-in-depth is the first fundamental principle. This principle proposes a multiple security layers approach. We can compare this foundation to a castle with the outer walls, inner sanctum, and final keep. The **principle of least privilege** is another keystone and forms a safeguard to ensure individuals have only the access they require. A breach at Target in 2013 demonstrates its cruciality. Initially, hackers accessed the network using credentials from an HVAC vendor, eventually moving laterally and compromising the payment system. Had the least privilege principle been properly applied, such lateral movement would have been far more challenging to realize.

Furthermore, it is paramount to ensure data integrity and confidentiality. We observed the fallout when this was not applied during the Equifax breach in 2017 when the personal data of millions of citizens was exposed. This example emphasizes the importance of both robust data protection measures as well as timely patching of known vulnerabilities.

The next key component, risk assessment, can be compared to a ship's radar, identifying potential obstacles. As a risk assessment identifies vulnerable systems and potential threats, the infamous Sony Pictures hack of 2014 might have been mitigated or even prevented with a thorough risk assessment.

Continuous monitoring and auditing have become non-negotiable. As an example, adequate monitoring could have detected unusual activity earlier, in J.P. Morgan's 2014 breach, when hackers accessed data for 83 million accounts, and it continued to be undetected for 2 months. Had this principle been applied, the breach's impact could have been drastically reduced.

Technical tools, from firewalls to encryption protocols, are like the shields and barriers of a fortress. But the human component – through training and awareness – is equally significant. The 2016 Podesta email leak serves as a stark reminder that proper training and awareness could have avoided the wide-reaching consequences of a simple phishing attack.

Emphasizing the importance of regular software updates and patches, we only need to recall the WannaCry attack's impact on systems running outdated Windows versions. Clear policies and guidelines, meanwhile, act as a compass, guiding behaviors. When companies have swift response policies potential damage can be minimized, such as the ones affected by the Cloudbleed bug in 2017.

In sum, a cybersecurity strategy is not just about having the right tools or tactics in place but rather about having an integrated approach where real-world lessons inform principles, processes, and actions. With a solid foundation, organizations can better navigate the tumultuous seas of cybersecurity, ensuring not only their survival but also their success.

Setting objectives and goals

Setting clear objectives and goals is the essence of cybersecurity. Crafting a resilient cybersecurity strategy is comparable to charting a voyage. The ship might be sturdy and the crew skilled, but without a clear destination and navigational plan, it's prone to drift. For example, for many organizations, a fundamental objective is to protect customer data. When looking back at the massive Marriott breach of 2018, where hackers accessed the personal details of approximately 500 million guests, a key takeaway was the emphasis on revisiting objectives, ensuring that customer data protection remains at the forefront.

Ensuring business continuity is another key objective. The 2017 Petya ransomware attack, which disrupted major companies such as Merck, Maersk, and FedEx, highlighted the need for organizations to prioritize continuous service provision even in the face of cyberattacks. These companies faced significant losses, with FedEx's subsidiary, TNT Express, alone facing an estimated $300 million in operational disruptions.

Furthermore, setting more specific goals within these objectives can drive more targeted actions. For instance, aiming to achieve a 95% patching rate within a 48-hour timeframe of a vulnerability disclosure can significantly reduce exposure windows. This was a lesson many organizations learned during the Heartbleed vulnerability of 2014, during which prompt patch application was crucial.

Moreover, as phishing remains one of the primary entry points for many cyberattacks, organizations could set a goal to train 100% of their staff annually in cybersecurity awareness. For example, when looking back at the 2016 attack on the Democratic National Committee, the risk of breach could have been drastically reduced with proper training.

Additionally, it remains invaluable to have objectives around regulatory compliance. British Airways, being fined $230 million in 2019 for GDPR violations following a data breach, underscores the importance of setting and achieving compliance-related goals. Not only will this mitigate financial risks, but it also preserves organizational reputation.

The faster an organization can detect and respond to a breach, the lesser the potential damage. Therefore, establishing goals around incident response times can be lifesaving. This can be demonstrated by the swift response during the 2019 Capital One Breach which helped limit the scope and fallout of the data leak.

Finally, metrics play a role as well as setting tangible targets can lead to more efficient security operations. For example, a target could be reducing false positives in threat detection by 50%. Regularly revisiting such metrics ensures that they remain relevant and drive continuous improvement.

To conclude, setting objectives and goals in cybersecurity consists of a strong mix of ambition, practicality, and adaptability, which are informed by lessons from the past. It provides the rudder and compass for the cybersecurity voyage, ensuring organizations do not just weather the storm but thrive in these challenging digital seas.

The role and significance of each element

Cybersecurity isn't just a monolithic construct; it's a dynamic tapestry of intertwined elements. Each component plays a distinct role, and together they form the bulwark against threats. Understanding the significance of each ensures a holistic and robust strategy.

As this can be seen as the "weather forecasting" of our cybersecurity journey, we start with risk assessment. Organizations can proactively defend themselves by identifying potential vulnerabilities. Stemming from an unpatched vulnerability, the Equifax breach in 2017 emphasizes the weight of thorough risk assessment and timely action.

When things go wrong, how quickly and effectively a company responds can make all the difference. Therefore, incident response plans are the contingency in operations. Maersk's handling of the 2017 NotPetya ransomware attack serves as an exemplar. Despite significant system disruptions, their swift and coordinated response ensured operational recovery in just ten days.

Security architectures and systems are the technical heart of cybersecurity and serve as our digital fortresses. They detect, deter, and defend. The 2013 Target breach, in which hackers exploited an HVAC system's weak security, highlights the need for comprehensive security architectures that leave no stone unturned.

Embodying the principle of least privilege, user access controls are the gatekeepers, determining who enters our digital realms and how far they can pass through. The 2014 Sony Pictures hack, where attackers roamed freely across the network, amplifies the need for robust access controls.

Security awareness programs are comparable to training the citizens of a fortress. They ensure that everyone from the sentry to the scribe, knows how to detect and repel invaders. This can be illustrated by the 2016 Podesta email leak, which resulted from a simple phishing attack, as this example drives home the importance of an educated and alert workforce.

Business continuity and disaster recovery plans are the blueprints for rebuilding post-breach. Having these plans in place ensures that operations can continue or swiftly resume, even after major incidents. The 2011 earthquake and tsunami in Japan provides a non-cyber analogy. Companies with robust recovery plans, such as Toyota, managed to bounce back faster than their peers.

As legal and regulatory compliance ensure operations are not just secure but also ethical and lawful, they are the charters and codes governing our digital domains. For instance, the GDPR fines imposed on British Airways in 2019 reiterate the need for a cybersecurity strategy in line with regulatory landscapes.

Lastly, continuous monitoring and improvement are the scouts and spies of our digital territories. Constant monitoring and improvement ensure a watchful eye on the horizon for new threats and inform the realm of necessary fortifications. Post-breach analyses highlight the essence of learning and evolving from each incident, the Capital One breach in 2019 highlights this.

To summarize, each element in a cybersecurity strategy carries distinct weight and significance. Recognizing the interconnectedness of elements, while informed by real-world incidents, ensures a rounded, resilient, and responsive strategy ready for the challenges of the digital frontier. Now, let's shift our focus to the core components of a cybersecurity strategy, the best practices for integrating them into the business model, and the potential challenges that might arise in the process.

Aligning cybersecurity strategy with business objectives

In today's rapidly evolving digital landscape, the intersection of business objectives and cybersecurity has never been more critical. At its core, cybersecurity is not just about protecting IT assets but about safeguarding the very essence of a business. Every organization, regardless of its niche, is fueled by its mission and vision. For example, when Adobe experienced a significant breach in 2013, the incident did not just compromise software; it jeopardized their customer's trust, which formed a cornerstone of their business model. As the infamous Yahoo! breach of 2013-2014 revealed, where data from all three billion user accounts was compromised, neglecting cybersecurity can dramatically derail business objectives. Cybersecurity is not a siloed endeavor; it's a holistic effort intricately tied to an organization's aspirations, reputation, and bottom line. Serving as another example, Cisco's 2018 breach, exposing a section of their infrastructure to potential malicious use, underlined how cybersecurity lapses can directly impact business functionality. The challenge for modern businesses is twofold: to prioritize cybersecurity initiatives that align with core objectives and to communicate the value of these initiatives to stakeholders. By achieving this alignment, organizations not only bolster their defenses but also enhance their competitive edge. Recognizing this synergy is the first step toward a more secure and prosperous future.

Correlation of organizational goals and cybersecurity endeavors

In the entangled world of modern business operations, cybersecurity cannot be an afterthought. The essence of any business's mission and vision revolves around offering value while ensuring trustworthiness and longevity. As cybersecurity plays a pivotal role in this promise, it should be intrinsically tied to the primary goals, vision, and mission of any organization. The case of the LinkedIn breach in 2012 exposing millions of passwords illustrates this. The incident did not just raise technical concerns but questioned LinkedIn's commitment to user security, impacting their overall brand trust.

Organizational endeavors carry inherent risks, whether they are service expansions, mergers, or digital transformations. For example, when Delta Airlines suffered a breach through a third-party chat service in 2018, it highlighted the necessity to scrutinize even peripheral systems in line with core business objectives.

Cybersecurity strategy must especially prioritize data integrity in sectors such as healthcare or finance, where the mission often involves handling sensitive information. The 2015 breach of health insurance giant Anthem, exposing the personal information of nearly 78.8 million members, serves as a strong example. The breach was not just about compromised data but rather a profound breach of the trust that members placed in the company to safeguard their intimate health and personal records.

Understanding the common relationship between strategic shift and cybersecurity is paramount for businesses undergoing digital transformation. For instance, when the retail giant Target decided to innovate its point-of-sale systems, the resultant 2013 breach affecting 41 million customer payment card accounts, underlined the importance of synchronizing business innovations with robust cybersecurity measures.

Often, an organization's mission and vision statements revolve around being market leaders, innovators, or trusted partners. Such objectives can only be genuinely realized when cybersecurity endeavors act as their foundation. Take the example of Sony Pictures in 2014. A devastating cyberattack not only disrupted their operations but also affected their reputation as a leader in the entertainment sector. It emphasized the intrinsic link between market leadership and cybersecurity resilience.

Besides, the integration of cybersecurity endeavors does not entail defense only but also enables businesses to seize opportunities confidently. For instance, companies such as Apple use their commitment to user privacy and security as a unique selling proposition, aligning their cybersecurity strategy with their broader business goals of user-centricity.

To summarize, the correlation between cybersecurity endeavors and organizational goals is not just a technical necessity but a strategic imperative for sustainable growth and building trust. Essentially, true success in the modern marketplace is not just about offering standout products or services but ensuring that they are delivered within a framework of trust, safety, and resilience.

Prioritizing cybersecurity based on business impact

In today's interconnected world, no business operates in isolation from cyber threats. The challenge lies in determining which threats to address first and how to allocate resources effectively. However, this challenge is magnified when considering the vast array of potential vulnerabilities and the diverse range of business models and objectives in play.

To illustrate this, the Equifax breach in 2017 serves as a stark example. In this breach, the personal data of nearly 143 million individuals was compromised, which not only resulted in a significant financial hit but also severely damaged the company's reputation. For a credit bureau, whose very foundation rests on trust and data integrity, this was a disastrous oversight. The lesson? Understand your business's cornerstone and prioritize its protection.

The essence of effective prioritization lies within the understanding of the potential business impact of a breach. This involves a thorough risk assessment that evaluates threats based on potential harm to business operations, financial impact, and reputational damage.

For instance, given that its entire business model revolves around cloud services, a SaaS-based startup might consider data breaches related to its cloud storage as its primary concern. On the contrary, a financial institution might prioritize defenses against phishing and financial fraud, given the direct and immediate implications of such attacks on its operations and clientele.

Beyond direct business implications, regulatory consequences also play a role. With GDPR coming into force in Europe, companies had to prioritize data protection and privacy to avoid massive fines. The growing weight of regulatory implications in the cybersecurity equation can be underscored by Facebook's $5 billion fine by the FTC over privacy violations in the Cambridge Analytica scandal.

Furthermore, businesses should factor in emerging trends and technologies as well. As the **Internet of Things** (**IoT**) expands and IT and IoT/OT converge, companies that produce smart devices may need to prioritize securing these devices due to the growing risks associated with them. Targeting IoT devices, the Mirai botnet attack in 2016 brought down vast portions of the internet, underlining the vulnerabilities in this area.

A pivotal strategy in prioritization involves the concept of "zero trust." This principle operates on the notion that no one, whether inside or outside the organization, is implicitly trusted. Adopting such an approach ensures that resources are allocated to protect against external threats and potential internal vulnerabilities as well.

Another vital approach is to engage in proactive threat hunting, rather than just building defenses. By actively seeking out potential vulnerabilities and addressing them before they are exploited, businesses can effectively mitigate risks.

In essence, prioritizing cybersecurity based on business impact requires a multi-faceted approach. Blending the elements of risk assessments, business objectives, regulatory landscapes, and emerging trends is like orchestrating a harmonious symphony from diverse musical notes. This multi-faceted approach should start with a clear understanding of the business's mission and vision, followed by mapping these objectives against potential threats through risk assessments. This landscape forms the baseline. Next, an overlay of regulatory requirements ensures that the strategy is compliant with external mandates. However, this is not a static picture. With the digital landscape rapidly evolving, it is essential to keep an eye on emerging trends, which could introduce new vulnerabilities or change the threat matrix altogether. Incorporating these into the strategy ensures it remains forward-facing. Only with a holistic view can organizations truly ensure that they are dedicating resources where they matter most, enabling them not just to defend but to thrive in the digital age. Regularly revisiting and updating this blended approach, through drills, audits, and feedback loops, will ensure that the cybersecurity strategy remains robust, relevant, and aligned with the business's overarching goals.

Communicating cybersecurity's value to stakeholders

For many businesses, cybersecurity remains a vague and ambiguous concept, often seen as the domain of IT professionals. However, as previously illustrated, its intrinsic value goes beyond technical defenses, directly impacting organizational trust, resilience, and growth. Effectively conveying this value to stakeholders, from board members to employees, is paramount to gaining necessary support and resources.

Conversations around cybersecurity often dive deep into technical jargon, which may alienate those not well-versed in the field. Therefore, start by tackling the language. Instead, translate technical cybersecurity metrics and risks into business terms. For instance, rather than discussing the complexities of a DDoS attack, articulate its potential to disrupt business operations and the financial implications of downtime. The **chief information security officer** (**CISO**) must be recognized as a business leader, not a technical leader. Real-world examples can be powerful tools in such conversations. The Yahoo! breach in 2013, which affected three billion accounts, led to a significant drop in the company's valuation during its sale to Verizon. This presents a clear example of how cybersecurity negligence can have a tangible financial impact.

Moreover, we must also understand the audience to effectively communicate value. A board member might be more interested in the long-term reputational risks and the potential financial impact of a breach, while a department head might be more focused on operational implications. Tailoring the message to your audience ensures resonance.

Additionally, visual aids, such as infographics or dashboards, can assist in making complex data more digestible. For example, a simple chart highlighting the increasing frequency of cyberattacks in a specific industry can help drive home the urgent need for investment in cybersecurity measures.

While the immediate cost of cybersecurity infrastructure might be evident, its long-term benefits, such as protection against potential fines or business continuity in the face of threats, might be less so. Therefore, it is essential to bridge the gap between cost and benefit. The case of British Airways, which faced a staggering $230 million fine for a data breach, showcases the value proposition of proactive investment in cybersecurity versus the reactive costs of non-compliance and breaches.

To add to the latter, it is important to demonstrate a **return on investment** (**ROI**). Especially at the executive level, stakeholders often look for ROI. While it might seem challenging to quantify cybersecurity ROI, you can approach it by calculating potential losses prevented, cost savings from avoided incidents, and the added value of maintaining customer trust.

Cybersecurity is not a one-off discussion; it is an ongoing conversation. Hence, stakeholders should be engaged through regular updates. By regularly sharing updates on threat landscapes, successful defenses, and areas of vulnerability, stakeholders remain informed and engaged.

Cybersecurity also aligns with ethical responsibilities. Emphasizing the moral obligation to protect customer and employee data can resonate deeply, especially in industries in which trust plays a pivotal role. The breach at health insurance provider Anthem, which compromised the personal information of 78.8 million members, is a stark reminder of the ethical implications of data protection.

To conclude, to be able to communicate cybersecurity's value effectively, you must uncover its complexities, align its importance with business outcomes, use tangible examples, and maintain an ongoing dialogue with stakeholders. In a world where cyber threats are ever-evolving, ensuring stakeholder buy-in, is not just strategic but essential for business longevity and trust. Moving forward, as we delve into the importance of incident response in the cybersecurity framework, it's crucial to remember that an effective strategy is not solely about risk management but also encompasses a comprehensive plan for when breaches occur. Hence, the spotlight turns toward intrusion detection and incident response systems.

Risk management and cybersecurity strategy

In the realm of cybersecurity, the saying *forewarned is forearmed* holds undeniable truth. Effective risk management is not just about preventing threats – it is about understanding the landscape, assessing potential impacts, and making informed decisions. The 2017 WannaCry ransomware attack, which affected over 200,000 computers across 150 countries, embodies the need for organizations to integrate risk management deeply into their cybersecurity strategies. While many organizations affected were caught off guard, those with a proactive risk management approach were better prepared to mitigate the impact of the attack.

Successful organizations do not just react to cyber threats – they anticipate and prepare for them. To build on this, they recognize that vulnerabilities are multiplex, and not all vulnerabilities require the same level of attention or resources. This necessary prioritization is achieved through comprehensive risk assessments, which not only identify threats but gauge their potential consequences and complications. To underscore this, we can take a look at the NotPetya malware outbreak, where companies that had previously assessed their vulnerabilities and had layered defenses in place managed the situation far better than their less-prepared counterparts.

Crafting a cybersecurity strategy without factoring in risk management is like navigating treacherous waters without a compass. This section examines the complexities of intertwining risk management with cybersecurity, guiding professionals on how to identify, assess, and prioritize risks effectively to ensure that their defenses are not just robust but also aligned with their organization's unique threat landscape and business objectives.

Integrating risk management methodologies in strategy formulation

As we cover the modern cybersecurity landscape, it becomes abundantly clear that mere reactionary measures are insufficient. Ensuring not just readiness but foresight in defense, we must strategically embed risk management methodologies within our cybersecurity strategies.

Understanding of the foundational principles of risk management lies at the heart of this integration. These principles go beyond the simple identification of threats. They encompass assessment, strategic mitigation, and diligent monitoring of risks. Additionally, when addressing cyber risks, it is pivotal to consider not only the technical aspects but also human elements, procedural complexities, and shifting external dynamics.

The critical role that existing frameworks and regulations play in shaping this integration cannot be emphasized strongly enough. Adopting and adapting recognized frameworks provides structure and direction. For instance, the NIST Cybersecurity Framework offers guidelines on risk management, tailoring them specifically to cyber threats. Its five core functions – Identify, Protect, Detect, Respond, and Recover – offer a holistic approach to understanding and managing cybersecurity risk.

Another globally recognized standard is ISO 27001, which focuses on establishing, maintaining, and improving an **information security management system (ISMS)**. By adhering to this framework, organizations can assure stakeholders of their commitment to safeguarding sensitive information. Many multinational corporations, such as Vodafone and IBM, have achieved ISO 27001 certification, showcasing their dedication to robust security practices and risk management.

For those in the financial sector, the FFIEC's guidelines provide industry-tailored insights. These guidelines emphasize the significance of risk assessment, strategy integration, and continuous monitoring in a financial context, in which security breaches can have catastrophic monetary and reputational consequences.

The importance of the integration of these frameworks into cybersecurity strategy is further supported by global and regional regulations. When the GDPR was rolled out, it was not just a data protection regulation but a loud call for businesses to reassess and redefine their cybersecurity strategies. The regulation emphasizes protecting user data, as non-compliance comes with hefty penalties. Likewise, California's CCPA underscores the importance of consumer data privacy, forcing businesses to integrate risk management practices that address these regulatory requirements.

However, merely adopting a framework or adhering to a regulation won't suffice. Organizations should internalize these guidelines and customize them so that they align with their specific business objectives. The 2017 Equifax breach shows that even though they were compliant with certain standards, a lack of holistic risk management led to a massive breach, impacting millions.

The breach poses an evident example of the pitfalls of generic, non-customized risk strategies.

If you're interested in digging deeper into these guidelines and standards, here are some public references:

- **NIST Cybersecurity Framework**: Visit the official NIST website for a comprehensive exploration of the framework and its implementation - https://www.nist.gov/cyberframework.

- **ISO 27001**: Access the ISO's webpage dedicated to ISO 27001 to understand the standard's structure, benefits, and certification process - https://www.iso.org/standard/27001.

- **FFIEC's guidelines**: The FFIEC IT Examination Handbook InfoBase provides a detailed look into the guidelines, including their applicability and implementation in the financial sector - `https://ithandbook.ffiec.gov/`.

- **GDPR**: The European Commission's Data Protection section provides a thorough understanding of GDPR, including its objectives, regulations, and enforcement - `https://commission.europa.eu/law/law-topic/data-protection_en`.

- **CCPA**: The California Department of Justice's CCPA section is a wealth of information on the act, its intent, and its implications for businesses - `https://oag.ca.gov/privacy/ccpa`.

Remember, it's not enough to simply know these frameworks and regulations. They should be actively integrated into your cybersecurity strategy, adapting to your specific business needs and objectives.

The process of integrating risk management methodologies extends beyond simple adoption. Strongly integrated risk management requires consistent engagement with stakeholders, commitment to continuous learning, and the agility to adapt to a constantly evolving digital threat landscape. Organizations can ensure a reinforced, comprehensive, and up-to-date approach to cybersecurity by placing frameworks and regulations at the core of risk management integration.

Conducting comprehensive risk assessments

Designed to proactively identify, evaluate, and prioritize vulnerabilities and threats, risk assessments are indispensable tools in the cybersecurity arsenal. Rather than simply reacting to breaches or threats as they arise, organizations must leverage risk assessments to systematically analyze their exposure to risk, anticipate challenges, and consequently tailor their defensive measures.

The following step-by-step guide provides the essential building blocks of efficient and effective risk assessment practice:

1. **Define risk assessment objectives and scope**:

    Before diving into the process, clearly outline what the main purpose of the assessment is and what you want to achieve. Are you just focusing on specific systems, on the entire organization, or maybe on new implementations? To ensure that there are no critical elements overlooked or left out in the risk assessment, a well-defined scope is of the essence. For instance, the 2013 Target breach was due to vulnerabilities in its third-party HVAC system, highlighting the importance of a wide-scope risk assessment that does not miss any secondary systems.

2. **Select the right tools and techniques**:

    The effectiveness of a risk assessment often relies on the available tools. From automated vulnerability scanners to manual penetration tests, each tool offers a unique perspective. To gain a comprehensive view it is important to apply a mix of tools. For example, while automated tools might detect an open port, a manual tester could exploit it, revealing the depth of a potential breach.

3. **Identify vulnerabilities and threats**:

    With the selected tools, span through systems, networks, and processes to identify vulnerabilities. Keep in mind that threats are not solely external. Insider threats, whether due to malicious intent or bare oversight, can be equally, and sometimes more, devastating. Edward Snowden and the NSA leak in 2013 stands as a strong reminder of the potential harm and consequences of insider threats.

4. **Evaluate potential impacts and likelihood**:

    It is important to note that not every vulnerability equates to an immediate or inevitable breach. Therefore, the potential impact of each identified vulnerability should be assessed, considering data sensitivity, system criticality, and operational consequences. Next, you must assess the likelihood of each threat occurring. To illustrate, a minor vulnerability in a non-essential system may be less concerning than a minor vulnerability in a core database.

5. **Assess current security controls**:

    This step is about evaluating the organization's current defense status. Are the current security controls adequate? Where are the gaps? For example, before the 2017 WannaCry ransomware attack occurred, many organizations believed they were secure, but those without updated patches were quickly proven wrong and faced the consequences.

6. **Document and analyze findings**:

    Documentation is not necessarily bureaucracy; rather, it functions as the foundation upon which strategic decisions are made. By systematically recording vulnerabilities, threats, impacts, and likelihoods, organizations can create a blueprint for action that serves as a point of reference for future assessments and strategic decisions.

7. **Prioritize remediation efforts**:

    Not all vulnerabilities demand immediate action. Based on the documented findings, prioritize vulnerabilities considering their impact and likelihood. Those that pose a significant threat to critical systems or data should be addressed directly. For instance, vulnerabilities that could expose customer data or compromise critical infrastructure should always be at the top of the prioritization list.

8. **Share findings and collaborate**:

    Risk assessments should not operate in isolation and findings should continuously be shared with key stakeholders across departments. Organizations should engage IT teams, management, and even end users when relevant to ensure a united approach in addressing vulnerabilities.

9. **Continuous monitoring and reassessment**:

    With new threats emerging daily, the cybersecurity landscape is fluid and ever-evolving, which requires periodic reassessment for organizations to stay ahead of potential threats. Based on technological changes, new information, and evolving threats, organizations should be regularly updating and adjusting their risk assessment practices.

In short, a comprehensive risk assessment is not a one-off task but an ongoing commitment, requiring a mix of tools, expertise, and collaboration. Organizations can strengthen their defenses and ensure agility and adaptability in a constantly evolving digital landscape by including comprehensive risk assessments in their cybersecurity framework.

Prioritization of mitigation strategies

In the large spectrum of cybersecurity, resources (time, money, or manpower) are scarce. It is crucial to prioritize mitigation strategies to ensure that the most critical vulnerabilities are addressed first and to allocate resources effectively and optimally.

The following step-by-step guide displays the essential building blocks of efficient and effective prioritization of mitigation strategies practice:

1. **Identify vulnerabilities and rank risks**: Conduct risk assessments and pinpoint vulnerabilities. Rank these vulnerabilities based on their potential impact and the likelihood of occurrence. High-impact, high-likelihood vulnerabilities demand immediate attention.

2. **Identify critical business functions**: Determine which vulnerabilities could potentially threaten your organization's critical business functions and designate these as high priority.

3. **Consider regulatory and compliance implications**: Some vulnerabilities may have legal or regulatory implications. Prioritize mitigation strategies that prevent non-compliance with regulations to avoid penalties under laws such as GDPR or HIPAA.

4. **Assess the financial impact**: Prioritize vulnerabilities that could directly affect your organization's financial stability. The goal is to minimize potential financial losses, both immediate and long-term.

5. **Evaluate stakeholder expectations**: Prioritize vulnerabilities that could significantly affect stakeholder trust and confidence. This includes customers, shareholders, and partners.

6. **Analyze the current threat landscape**: Stay up-to-date with the ever-evolving threat landscape. If a widespread attack targeting a specific vulnerability is happening, accelerate mitigation strategies for that particular vulnerability.

7. **Conduct remediation cost versus impact analysis**: Weigh the cost of remediation against the potential impact. It may be more cost-effective to implement an interim control for a low-impact vulnerability while allocating more resources to a high-impact one.

8. **Evaluate technological feasibility**: Consider the technological implications of the mitigation strategies. Ensure that the chosen strategy aligns with your organization's technological capabilities.

9. **Establish a feedback loop**: Implement a feedback mechanism to continuously assess the effectiveness of strategies. Make timely adjustments as needed.

10. **Implement a multi-layered defense approach**: Consider how each mitigation strategy fits within the broader defense framework to ensure resilience even if one layer is compromised.

11. **Encourage a collaborative approach**: Foster collaboration among different departments to align mitigation strategies with business goals and processes.

12. **Schedule periodic re-evaluation**: Regularly revisit and re-prioritize mitigation strategies to adapt to the evolving threat landscape and organizational changes.

To conclude, prioritization of mitigation strategies is not a static or fixed task but rather involves a continuous process of evaluation and re-evaluation. It requires a nuanced and full understanding of the organization's landscape, stakeholder expectations, and the wider threat landscape. Having an adequate mitigation strategy in place enables organizations to realize a robust and agile defense mechanism, countering threats and minimizing impacts. With the foundational understanding of incident response planning and preparedness firmly established, let's now transition to the following pivotal stage. This section will delve into the mechanics of incident detection and analysis, a crucial component that effectively links planning to decisive action in the cybersecurity landscape.

Incident response planning and preparedness

In this section, we'll cover incident response planning and preparedness. We'll explore the precise and accurate building of procedures, structure the incident life cycle, and cover the interdependent role of tools and human expertise.

As security incidents are unavoidable, a well-prepared incident response plan is essential. Potential damage and impact of breaches can be significantly mitigated by swift and strategic action when responding to incidents.

A good incident response plan entails more than technical solutions alone – it involves clear procedures, efficient communication, and a deep understanding of incident life cycles.

First of all, the design of tailored incident response procedures is essential. A deep understanding of an organization's potential threat landscape and infrastructure is important to be able to identify potential security events. Additionally, the protocols in place should be unambiguous, clearly describing the responsibilities and the actions that should be taken throughout the response process. Simulated breaches and tabletop exercises can help test the procedure's effectiveness as well as highlight its value.

Incident management and response should be applied with a structured process, starting from the detection phase and culminating in the post-incident analysis.

Detection requires cautious monitoring and surveillance, which can be realized by tools detecting anomalies, irregular patterns, and unauthorized access. Moving forward, the containment and elimination phase involves balancing between immediate short-term measures and enduring elimination of threats. The Ukraine power grid attack in 2015 exemplifies this phase, by demonstrating the drastic consequences of inadequate containment. Recovery follows, which focuses on the recovery and reformation of regular operations.

Additionally, the human element is a foundational pillar in successful incident response. While essential technology tools include incident detection software and analytics platforms, the true heroes are the members of the incident response team. Therefore, it is essential that the incident response team is carefully selected and trained and that members have clearly defined roles. Seamless communication and collaboration within the team, as well as across departments, are crucial factors for success. The 2015 United Airlines incident serves as an exemplifier here as it highlights the significance of quick and effective inter-departmental communication when a cybersecurity expert discovered vulnerabilities in the in-flight entertainment system, which mitigated potential threats.

Designing tailored incident response procedures

Being capable of responding promptly and effectively to incidents is crucial to realizing a resilient defense strategy for an organization. To be relevant and effective, incident response procedures should be tailored to the organization's operational realities and unique threat landscape.

The following list covers the essential elements of designing tailored incident response procedures:

1. **Identifying potential security events**: Firstly, a comprehensive identification of potential security events should be realized. This requires a profound understanding of the organization's digital and physical infrastructure. Organizations can gain real-time visibility into network activities and potential anomalies by leveraging technologies such as **security information and event management** (SIEM) systems. This can empower organizations to recognize patterns that possibly indicate security incidents, fostering a proactive approach to incident response planning. Moreover, threat intelligence platforms and feeds could offer insights into industry-specific emerging threats, supporting the identification of potential security events that require organizational preparedness.

2. **Defining clear protocols**: Once potential security events have been identified, the next step is to define clear protocols for each stage of incident response. The protocols should serve as a framework for defining roles, responsibilities, and actions, as well as contain communication strategies for both internal teams and external stakeholders. Streamlining communication and coordination during an incident can be facilitated by vital technologies such as **security orchestration, automation, and response** (SOAR) platforms. These tools can automate repetitive tasks, trigger predefined response actions, and facilitate real-time collaboration in incident response teams. Moreover, standardizing responses and minimizing decision-making time in high-stress situations are supported by so-called incident response playbooks. Good playbooks will structure step-by-step guides on procedures and expected actions to be carried out during different scenarios.

3. **Simulation and testing**: Simulated breaches and tabletop exercises are powerful tools for testing and evaluating an organization's incident response procedures and practices. These exercises consist of realistic threat scenarios requiring the response team to navigate them. Organizations can leverage cyber range platform technologies to simulate cyberattacks in controlled environments, enabling the response teams to practice their skills and procedures. Furthermore, organizations can monitor how incidents unfold across endpoints and networks by leveraging **endpoint detection and response (EDR)** and **extended detection and response (XDR)** tools. These technologies can help provide insights into the effectiveness of incident response procedures as well. Organizations can fine-tune their response strategies based on the actual behaviors exhibited during the simulated incidents.

4. **Documenting and iterating**: To create comprehensive playbooks, effective incident response procedures should be precisely shared and documented. These playbooks can then serve as the main reference point for response teams during high-pressure situations, ensuring consistent actions are taken. Furthermore, quick access to critical information can be granted by integration documentation platforms with incident response tools and dashboards.

    Finally, it is important to note that incident response procedures should not be static. The playbook must be repeatedly re-evaluated to ensure the procedures stay effective and relevant. Procedures should be regularly updated to effectively address emerging threats by monitoring the evolving threat landscapes and technological advancements.

5. **Tailoring for the human element**: Incident response procedures must cater not only to technical aspects but also to the human element. A critical aid to understanding the human element can be the use of **user and entity behavior analytics (UEBA)** solutions, which enable organizations to analyze user behavior and detect anomalies of insider threats or compromised accounts. UEBA-type technologies aid in understanding user patterns and identifying anomalies that might require immediate action. Additionally, security awareness and training programs play an important role in preparing teams for incident response. By educating employees about cybersecurity best practices and response protocols, organizations empower them to act swiftly and decisively in the face of an incident.

To summarize, designing tailored incident response procedures requires a balanced application of technological tools, strategic planning, and people-focused considerations. By leveraging technologies such as EDR and XDR tools, organizations can streamline coordination of incident detection and response as well as documentation of best practices and procedures. With constant re-evaluation, refinement, and adaptation, these procedures enable organizations to respond effectively to evolving cyber threats, minimizing damage and maintaining operational resilience.

The incident management life cycle

Understanding the life cycle of incident management can be overwhelming and sometimes confusing because depending on where you read about it, you might find different explanations. But at its core, it's about detection, containment, eradication, recovery, and reflection of an incident. So, let's go through each of these steps:

1. **Detection**: Detection is the gateway to incident response. Technologies such as **identity threat detection response (ITDR), EDR, intrusion detection systems (IDSs)**, and **intrusion prevention systems (IPSs)** serve as sentinels, scanning the enterprise for anomalies. Many organizations often pair these technologies with UEBA which adds a layer of intelligence, detecting deviations from normal user behavior. These technologies, complemented by continuous monitoring and data correlation, elevate the chances of swift detection.

2. **Containment and eradication**: Once a threat has been detected, containment and eradication are immediate next steps. This phase requires tactical maneuvers and strategic foresight. Isolating compromised systems from the enterprise network curtails the spread of the threat. Leveraging technologies like Network Segmentation ensures that a breach in one segment doesn't infiltrate the entire network. Simultaneously EDR and XDR solutions facilitate real-time monitoring and enable the identification of malicious processes and behavior.

3. **Recovery**: Recovery is the bridge between the chaos of an incident and the restoration of business operations. This phase demands a balance of speed and precision. Leveraging automated backup and disaster recovery solutions ensures data availability and minimizes downtime. Technologies such as virtual machine snapshots allow for rapid system restoration. Cloud-based solutions further expedite recovery, enabling seamless data retrieval even if on-premises systems are compromised.

4. **Post-incident analysis and review**: A step that is often forgotten due to the sheer volume of activities. Post-incident analysis involves dissecting the incident's anatomy to understand its origins, implications, and modus operandi. Technologies such as SIEM systems aggregate and correlate data, aiding in reconstructing the incident's timeline. The analysis serves as a blueprint for refining incident response procedures, addressing vulnerabilities, and enhancing future defenses.

5. **Incident response frameworks**: The life cycle of incident management is fortified by the structure of incident response frameworks. Frameworks such as the NIST Cybersecurity Framework and the MITRE ATT&CK Framework provide systematic guidelines for organizations to approach incident management. The NIST Framework, for instance, consists of functions – Identify, Protect, Detect, Respond, and Recover – that provide a comprehensive roadmap for managing incidents. The MITRE ATT&CK Framework delves deeper into threat tactics, techniques, and procedures, empowering organizations to proactively defend against known attack patterns. By embracing these frameworks, organizations ensure a holistic approach, spanning detection, analysis, response, and recovery. They guide organizations in building proactive defenses, minimizing incident impact, and enhancing response efficiency. For example, the adoption of the NIST Cybersecurity Framework by the US government and various industries reflects its comprehensive approach to incident response.

Organizations that strategically employ detection technologies, practice swift containment and eradication, embrace resilient recovery methods, and ardently analyze post-incident scenarios are better poised to weather the storm of cyber threats. Through continuous refinement and learning, the orchestration of this life cycle transforms into a virtuoso performance of cybersecurity resilience.

Tools, technologies, and human elements in incident response

Effective incident response goes beyond just the technology but also factors in human expertise. In this chapter, we will overview the fusion between tools, technologies, and people that enables effective incident response. In the ensuing sections, we will delve further into the essential components of an effective incident response strategy, meticulously dissecting every facet:

- **Critical tools and technologies**: Incident response thrives on an arsenal of tools and technologies that enable them to perform quick actions. Among these, SIEM systems often are leveraged as data aggregators and analyzers, providing insights into enterprise-wide activities. By integrating threat intelligence feeds, SIEMs can correlate data with known threat indicators, heightening the chances of early detection. EDR, ITDR, or **network detection and response** (**NDR**) solutions are often leveraged for post-breach detection and response workflows on their protected surfaces. The rise of XDR solutions takes this a step further by correlating data across multiple surfaces and third-party data.

- **The role of automation and playbooks**: Incident response management provides ample opportunities for automation and orchestration. By integrating SOAR systems, firms can mechanize routine processes, expedite response rates, and make workflows more efficient. Using automated playbooks, specific actions can be set off based on certain triggers, thereby hastening isolation and elimination. Integrating these platforms with threat intelligence feeds ensures that measures taken are guided by up-to-the-minute threat data. This blend of human skill and automated responses not only enhances response efficiency but also assures uniformity and minimizes human error.

- **The role of the incident response team**: At the heart of effective incident response is the incident response team – a group of experts collaborating to engage against cyber threats. Your incident response team should comprise roles such as incident commander, forensics analyst, communications lead, and legal advisor, each with defined responsibilities. Continuous training, workshops, and scenario-based exercises not only hone individual skills but also foster cohesion within the team. Teams can be likened to Special Forces units – trained and ready to execute precise actions when called upon.

- **Human factor**: While emphasizing technology, the human aspect remains crucial. The psychological effect on those responding to incidents is of great importance. High-pressure situations require a robust mindset. Those in charge of incident response must engender a culture that accentuates self-care, stress alleviation, and mutual support. Post-incident briefings facilitate a platform for team members to convey their experiences and learn from one another. In the domain of cybersecurity, organizations must recognize the emotional strain and cultivate an environment that enables teams to excel even under intense pressure.

- **Collaboration and communication**: Teams engaged in incident response need to interact flawlessly among team members and various departments. Advanced tools such as incident management platforms enable immediate information exchange, letting teams synchronize activities, monitor progression, and disseminate crucial updates. The dialogue must not be limited to the response team alone – stakeholders, top executives, legal advisors, and public relations teams need to be updated promptly to ensure a coherent and knowledgeable approach.

The cogent response to cyber incidents is an endeavor that requires a multitude of tools and human expertise woven together comprehensively. The ability to effectively respond to cyber threats combines several interconnected processes functioning in harmony. This includes the use of tools to detect and respond to threats, the formation of highly skilled incident response groups with the acumen to handle complex cyberattacks, the cultivation of open and efficient communication channels within the organization, and the integration of automation to expedite detection and response processes. Moreover, recognizing and addressing the psychological aspect of cybersecurity – the human factor – is crucial. Next, we'll let's focus on the foundation of understanding the importance of the human factor and security awareness training.

Security awareness and training programs

The critical nature of comprehensive security precautions is highlighted by the ever-increasing threat landscape. Cybersecurity is no longer only the responsibility of the security team; instead, it is a collective obligation spread throughout every business level. In this chapter, we will focus on the crucial function of security awareness and training programs in fostering a vigilant environment and arming staff with the tools to defend against ever-changing cyberattacks by embracing a security-centric methodology.

As companies increasingly rely on interconnected systems and undergo digital transformation, the human factor becomes a vital component in the cybersecurity defense chain. This is why it's often referred to that humans are the weakest link in the enterprise. Cybercriminals frequently manipulate humans by exploiting their weaknesses via strategies such as phishing and social engineering. Without employee readiness to identify and react to these threats, even the most cutting-edge technical defenses can be undermined. This underlines the critical importance of comprehensive security awareness and training programs.

Role-based enablement is a fundamental element of effective security awareness initiatives. Acknowledging that each role within an organization carries specific responsibilities and has unique potential weak points, these training programs cater to different job functions. Whether an employee is in charge of confidential revenue predictions or participates in software development, targeted training modules ensure the shared information is pertinent and actionable. This individualized approach maximizes engagement and encourages employees to apply security best practices directly to their tasks.

It is important to note that the training efficiency should not be presumed – it needs to be evaluated and consistently enhanced. Regular assessments, such as simulated phishing attacks and post-training assessments, generate insights into the knowledge retention and response capabilities of employees. These metrics not only guide training improvements but also produce concrete evidence of the program's effect on risk reduction.

Security awareness and training programs aim to equip employees with the skills and knowledge they need and foster a security-first mindset. This culture shift involves comprehending security principles and integrating them into the organizational DNA. By incorporating security values into their core mission and values, companies create an environment where security-conscious behavior is celebrated and advocated.

In this chapter, we will dissect the strategic components of creating and executing efficient security awareness and training programs. From customizing training content to assessing its impact and fostering a culture of ongoing learning, each facet contributes to an organization's collective defense against cyber threats. Much like a well-practiced orchestra creates a harmonious symphony, a well-structured security awareness program nurtures a workforce united in vigilance and readiness.

Tailored training for organizational roles

In the intricate tapestry of cybersecurity, knowledge is a potent shield. Yet, the effectiveness of this shield is not solely determined by its strength but by how well it aligns with an individual's responsibilities. This is why role-specific education forms the cornerstone of a resilient defense against the threat landscape. In this section, we will dive deeper into what it takes to create training programs that resonate with different roles and explore how organizations can achieve this strategic imperative. To effectively mitigate cybersecurity threats across all levels of an organization, it is crucial to adopt a role-centric approach to security training. The following factors shed light on the importance and implementation of role-specific training in cybersecurity:

- **The importance of role-specific training**: Cybersecurity is not a one-size-fits-all endeavor. Different roles within an organization are associated with distinct responsibilities, access levels, and potential vulnerabilities. As a simple example, the level of training required for the security team will be naturally different to the marketing team which again would be different to the finance team. Tailoring training to specific roles ensures that the acquired knowledge is directly applicable, maximizing its impact. For example, while finance personnel need to be adept at recognizing spear-phishing attempts targeting financial transactions, developers should focus on secure coding practices to prevent vulnerabilities in software. Recognizing these nuances empowers organizations to offer targeted training that guards against role-specific threats.

- **Leveraging learning management systems (LMSs)**: The orchestration of tailored training is facilitated by LMS. These platforms streamline the creation, delivery, and tracking of training content. LMS allows organizations to structure training materials according to their roles, making them accessible to the right individuals. For example, finance staff can access modules related to financial fraud prevention, while IT administrators can delve into network security. LMS platforms also enable organizations to monitor employee progress, identify knowledge gaps, and ensure compliance with training requirements.

- **Aligning training with job functions**: An important step in achieving role-specific training excellence is aligning training content with specific job functions. This alignment ensures that training modules are not just educational but actionable. Infusing scenarios and case studies that mirror real-world situations faced by individuals in their roles fosters a deeper understanding of cybersecurity principles. At the end of the day, you want to ensure the content resonates and your key messages land. For instance, a customer service representative's training could include handling customer inquiries about potential phishing emails.

- **Real-world success**: Microsoft's approach to role-based security training serves as a great example model. Recognizing the diversity of roles within their organization, Microsoft offers a range of role-specific training modules. Their security awareness training program provides tailored content for engineers, sales professionals, and executives, each focusing on the unique security challenges inherent to their roles. This approach enhances engagement and ensures that employees possess the skills to combat threats relevant to their responsibilities.

In conclusion, the effectiveness of cybersecurity awareness hinges on its relevance to individual roles. Crafting tailored training that aligns with organizational responsibilities empowers employees with the knowledge and skills to safeguard against role-specific threats. By identifying unique challenges, leveraging learning management systems, and weaving real-world scenarios into the content, organizations can seamlessly infuse cybersecurity into daily operations. Just as a skilled archer aligns their shot with precision, organizations can align tailored training with individual roles, hitting the target of cybersecurity excellence.

Continuous evaluation and improvement

As organizations invest in security awareness and training programs, a critical question arises: how do they ensure that these programs remain effective over time? Knowledge, especially in cybersecurity, is not static but requires constant evaluation, adaptation, and improvement. Therefore, the answer lies in the realm of continuous evaluation and improvement – an iterative process that sharpens the edge of cybersecurity education. In this section, we will delve into the different components of assessing training effectiveness and iterating to address evolving knowledge gaps, along with actionable insights on how to achieve this ongoing enhancement.

Continuous evaluation and improvement start with a robust and effective evaluation model. As an example, if you were to continue delivering the same security awareness training every time, at one point, the content will become irrelevant and you will lose the interest of the participants. This is why merely delivering training modules isn't sufficient; organizations need to gauge their impact. One metric of assessment is the click-through rates in simulated enablement exercises such as a phishing simulation. In the example of a phishing simulation, the goal is to mirror real-world phishing attacks and measure how effectively employees can identify and avoid potential threats. For instance, if an organization notices a decline in click-through rates after a training module, it's an encouraging sign that the training has bolstered employees' ability to recognize phishing attempts.

Following the awareness and training sessions, it's important to step back, reflect, and improve. To achieve that post-training assessments for retention are critical. It's not enough for employees to grasp security concepts momentarily – they must retain and apply this knowledge. Post-training assessments play a pivotal role in evaluating knowledge retention. After completing a training module, employees take an assessment that tests their understanding of the material. These assessments provide insights into the extent to which employees have picked up the newly learned security practices. For example, a marketing professional who received training in detecting marketing scams should be able to identify potential threats in a post-training assessment scenario.

We all have attended various trainings in our career. Stall training, where we get loaded by presentations and readouts, can be challenging to follow and worse to remember. This is why good security awareness training includes iterative enhancements for the training content. The beauty of continuous evaluation lies in its transformative potential. As assessments reveal knowledge gaps and areas for improvement, organizations can iteratively enhance training content. If an assessment highlights that employees struggle with identifying sophisticated phishing emails, the organization can introduce more advanced scenarios in its training modules. Iterative enhancements ensure that training content remains relevant and impactful as cyber threats evolve.

There are so many different ways to measure the success of a security awareness training program. The fundamental question it must address is whether we contributed to developing a security-first mindset. Therefore, actionable insights for achievements are critical. Achieving continuous evaluation and improvement necessitates a structured approach. Begin by defining **key performance indicators (KPIs)** that align with training objectives. These KPIs could include a reduction in click-through rates, an improvement in assessment scores, and increased reporting of suspicious activities. Continuously monitor and analyze these metrics to identify trends and patterns. When discrepancies arise – such as a sudden spike in phishing-related incidents – investigate root causes and adjust training content accordingly.

The **National Institute of Standards and Technology** (**NIST**) offers a real-world example of continuous evaluation and improvement in practice. NIST's Phish Scale project regularly assesses employees' susceptibility to phishing attacks using a controlled testing environment. The results of these assessments guide the refinement of training content and strategies to mitigate phishing risks. This iterative approach has led to a significant reduction in click-through rates and enhanced overall cybersecurity awareness.

In the dynamic landscape of cybersecurity, complacency is the adversary. Continuous evaluation and improvement stand as sentinels guarding against stagnation. By measuring training effectiveness, assessing knowledge retention, and embracing iterative enhancement, organizations remain adaptive in the face of evolving threats. Just as a vigilant sentry patrols the walls of a fortress, continuous evaluation ensures that the walls of cybersecurity remain fortified against an ever-changing threat landscape.

Fostering a security-first mindset

Technical measurements are only as strong as the human element that uploads them. Therefore it is critical to learn and adapt a security-first mindset not only within the security department but across the company. A security-first mindset is a beacon that guides organizations from mere compliance to a transformative cultural shift – a shift where cybersecurity isn't just a set of protocols, but a collective ethos that permeates every facet of the organization.

In this section, we will dive deep into embedding security principles, creating an ecosystem where cybersecurity consciousness thrives, and people are encouraged to take ownership and report suspicious behavior without fear or retaliation.

A security-first mindset goes beyond protocols and procedures but is a fundamental paradigm shift for employees as vigilant guardians of enterprise data and services. Adapting a security-first approach goes beyond just the simple reporting of possible phishing emails but instead generates a sense of belonging and responsibility within each employee. Achieving this level of maturity is hard and not achieved overnight and requires weaving cybersecurity awareness into the fabric of organizational culture.

A key component of this is embedding security into the core values and mission of the organization. Values and mission statements serve as the compass that guides an organization's journey. By embedding cybersecurity principles into these foundational elements, organizations send a clear message not just to their employees but also to business partners, suppliers, and customers that security isn't a separate concern – it's integral to the organization's identity. For example, an organization's value of "integrity" could extend to maintaining the integrity of data and systems through robust security practices.

Cybersecurity is constantly evolving, so we must encourage continuous learning through active engagement. An engaged workforce is the cornerstone of a security-first culture. To achieve that, it is important to promote continuous learning through gamified training modules, newsletters, and lunch-and-learns. Gamified modules turn cybersecurity education into an interactive adventure, fostering competition and engagement. Regular newsletters disseminate insights on emerging threats and best practices. Lunch-and-learns provide a platform for knowledge sharing and discussions, creating an atmosphere of curiosity.

As with many things, leadership can be a catalyst, and adopting a security-first mindset should be led by example. The involvement of senior leadership is pivotal in fostering a security-first mindset. When leaders champion security awareness, it sends a strong message that cybersecurity isn't relegated to the IT or security department – it's an organizational priority. Leadership engagement can take the form of regular communication, participation in enablement sessions, and leading by example in adhering to security practices.

Google's **Phish Alert** button exemplifies the power of a security-first mindset within the workforce. This internal tool empowers employees to report suspicious emails with a single click. The effectiveness of this tool not only lies in its technical utility but also in the cultural message it conveys – every employee is a frontline defender against the ever-evolving threat landscape. This collective vigilance has contributed to Google's robust security posture.

In the realm of security awareness and training programs, the true measure of success is a workforce that's not just informed, but empowered – a workforce that instinctively acts as a bulwark against cyber threats. By fostering a security-first mindset, organizations rewrite the script from reactive defense to proactive vigilance. Just as a symphony reaches its crescendo when every musician is attuned to the music, organizations reach their cybersecurity pinnacle when every employee is attuned to the symphony of security.

Summary

In this chapter, we delved into the complexities of creating effective cybersecurity measures in a constantly evolving digital landscape. It critiques traditional approaches such as the "moat and castle" strategy for being inadequate against sophisticated cyber threats, emphasizing the need to integrate cybersecurity with core business goals. We underscored the importance of risk management, making it a central part of strategic planning rather than a peripheral consideration. It discusses the critical roles of digital forensics and incident response in modern cybersecurity environments and highlights the necessity of security awareness across all team members. By doing so, it aims to foster a deeper understanding of the challenges and opportunities in cybersecurity.

Furthermore, this chapter outlined the key elements of a successful cybersecurity strategy, which includes aligning it with business objectives, focusing on risk management, planning for incident response, and developing security awareness and training programs. This chapter argued that a strong cybersecurity strategy requires a balance between risk, business objectives, and the dynamic nature of cyber threats, rather than relying solely on defensive mechanisms or advanced technology. Real-world examples, such as the WannaCry ransomware attack and the Equifax data breach, are used to illustrate the consequences of inadequate cybersecurity measures and the importance of robust strategies. This chapter concluded by emphasizing the significance of an integrated approach to cybersecurity, where real-world lessons inform principles, processes, and actions, creating a solid foundation for organizations to navigate the challenges of the digital age.

6
Aligning Security Measures with Business Objectives

Ensuring that corporate cybersecurity initiatives are aligned with clear business objectives has become imperative. This is foundational to ensuring that security measures contribute positively to business value. The ability to discern this alignment elucidates cybersecurity's integral role in navigating an organization toward its strategic milestones.

Identifying and prioritizing security initiatives based on their potential impact and associated risks to business operations are crucial for effective resource allocation. This prioritization ensures that efforts and investments in cybersecurity are channeled toward areas with the most significant business impact. Acquiring the competency to prioritize security initiatives that resonate with business priorities optimizes the risk management process, facilitating a proactive rather than reactive cybersecurity posture.

To achieve this, it is paramount that we understand how to articulate security initiatives and investments to non-technical stakeholders. However, often, this very thing can be a challenging task. Effective communication is critical to garnering organizational buy-in and ensuring sustained investment in cybersecurity. Strategies and frameworks that help communicate security initiatives' **return on investment (ROI)** effectively elucidate how these investments mitigate risks and enable **business continuity (BC)**. This engagement fosters a culture of shared responsibility toward cybersecurity, thereby promoting an organization-wide, security-first mindset.

This chapter aims to bridge the often-observed chasm between the business and cybersecurity realms, promoting a culture where cybersecurity is perceived and operated as a business enabler. By diligently exploring these areas, readers will be better positioned to drive a security strategy that mitigates risks and propels business objectives forward. The alignment of security measures with business objectives is a pivotal stride toward nurturing a resilient, adaptive, and growth-centric organizational ethos.

This chapter will cover the following topics:

- The importance of aligning security with business objectives
- Prioritizing security initiatives based on risk and business impact
- Communicating the value of security investments

Let us dive in!

The importance of aligning security with business objectives

Aligning security measures with business objectives has transitioned from a technical requisite to a strategic necessity, and those organizations that can pivot are those that set up their cybersecurity strategy for success and can safeguard the organization. The escalating complexity of cyber threats and regulatory mandates necessitates a seamless integration between cybersecurity initiatives and overarching business goals.

We will explore instances underscoring the substantial impact of well-aligned cybersecurity measures on business performance. The discussion extends to establishing a link between business objectives and security measures, an effort that calls for active engagement from key stakeholders. This sets the stage for a profound understanding of the symbiosis between cybersecurity and business objectives, laying the foundation for fostering a proactive, business-centric cybersecurity posture.

The critical role of cybersecurity in business environments

The essence of business operations has been entwined with frameworks, elevating cybersecurity's importance from a mere technical requirement to a strategic business enabler. As organizations continue to undergo digital transformation, the traditional perspective of viewing cybersecurity merely as a siloed IT function has radically transformed and profoundly reimagined how cybersecurity can act as a catalyst in achieving overarching business objectives.

Misalignment between security and business objectives can result in adverse repercussions, not just regarding data breaches or compliance failures, but in a broader scope, affecting market position, customer trust, and financial health. The fallout from a misaligned cybersecurity strategy can echo through organizational corridors long after the initial incident, imprinting a lasting negative impact.

The discrepancies arising from such misalignment can be seen in instances where the lack of a robust, aligned cybersecurity posture significantly impeded business performance. For example, data breaches resulting from misaligned cybersecurity measures have caused numerous organizations substantial financial and reputational damage. Conversely, organizations tend to thrive even in a hostile digital environment when cybersecurity measures are meticulously aligned with business objectives. For instance, organizations that have invested in robust cybersecurity frameworks, aligning them with their business goals of trust and reliability, have witnessed enhanced customer loyalty and trust. This alignment has fortified their security posture and created a competitive advantage in the fiercely competitive financial market.

Additionally, enterprises in the healthcare sector, which have prioritized cybersecurity alignment with their business objective of safeguarding patient data, have seen fewer data breaches, thereby upholding their reputation and ensuring regulatory compliance. This alignment has further facilitated digital transformation initiatives within these enterprises, propelling them toward enhanced operational efficiency and improved patient care.

In a global market where competition is fierce and the threat landscape is continuously evolving, aligning cybersecurity measures with business objectives is no longer optional; it's imperative. Cybersecurity is not an isolated technical domain but a strategic business enabler that can significantly impact an organization's trajectory toward its goals.

The journey from viewing cybersecurity as a technical requirement to acknowledging it as a strategic business enabler is laden with challenges and learning curves. Yet, it is a journey that modern organizations must undertake to navigate the intricate maze of today's digital business environment securely and efficiently. Organizations are better positioned to mitigate risks, seize market opportunities, and drive sustainable business growth in an increasingly digital world through a diligent, well-thought-out alignment of cybersecurity initiatives with business objectives.

Connecting business objectives and security measures successfully

A pivotal step in this alignment is identifying vital organizational stakeholders who will play instrumental roles in bridging the gap between cybersecurity initiatives and business goals. In order to construct an environment conducive to alignment between cybersecurity initiatives and business objectives, we must take a systematic three-pronged approach: identifying key stakeholders, defining a common language, and developing an integrated security and business strategy framework. Let's look at this in more detail:

- **Identifying key stakeholders**: The journey toward alignment begins with identifying key stakeholders who hold interests in the cybersecurity posture and the organization's business performance. These stakeholders often span various organizational tiers, including executive leadership, board members, **business unit** (**BU**) leaders, and the cybersecurity team. Their collective expertise and decision-making capabilities are essential in fostering a cohesive approach toward aligning cybersecurity measures with business objectives. For instance, executive leadership and board members provide strategic direction and resources, BU leaders offer operational insight, and the cybersecurity team provides the technical expertise required to navigate the complex cybersecurity landscape.

- **Defining a common language**: Establishing a common language that translates the technical jargon of cybersecurity and resonates with the business lexicon is fundamental in facilitating meaningful discussions and collaborative efforts. This common language helps articulate the value and implications of cybersecurity initiatives in a manner that is understandable and relevant to non-technical stakeholders. It aids in translating technical risks into business risks and security investments into business value, enabling informed decision-making.

- **Developing an integrated security and business strategy framework**: Aligning security measures with business objectives necessitates a structured approach, which can be achieved by developing an integrated security and business strategy framework. This framework serves as a blueprint that guides the alignment process, ensuring that cybersecurity initiatives are meticulously mapped to business goals. It outlines the strategic, operational, and technical measures required to safeguard organizational assets while propelling the business toward its objectives.

One pivotal component of this framework is conducting a thorough risk assessment to understand cybersecurity threats and vulnerabilities. This risk assessment aids in prioritizing security initiatives based on the potential impact on business objectives. Subsequently, developing a cybersecurity strategy that is intertwined with business strategy becomes more achievable. It encompasses setting clear goals, defining roles and responsibilities, and establishing metrics to measure the effectiveness and impact of cybersecurity initiatives on business performance.

Moreover, this framework should encourage continuous communication and collaboration between cybersecurity and business stakeholders. It must promote a culture of shared accountability toward achieving a robust security posture that enables business success. The framework remains dynamic and adaptable to the evolving threat landscape and changing business objectives through regular reviews and updates.

By identifying the right stakeholders, establishing a common language, and developing an integrated security and business strategy framework, organizations lay a strong foundation for aligning cybersecurity measures with business objectives. This alignment is not a one-time effort but an ongoing process that requires steadfast commitment, effective communication, and collaborative efforts across the organizational spectrum.

In conclusion, aligning cybersecurity measures with business objectives is a holistic approach encompassing strategic planning, operational execution, and continuous improvement. It requires a concerted effort from all stakeholders, united by a common understanding and a shared vision toward achieving robust cybersecurity and business success. Through this alignment, organizations are better poised to navigate the intricate digital landscape securely and efficiently, ensuring sustainable business growth in an increasingly connected and competitive global market.

Measuring the impact and value of aligned cybersecurity initiatives

The alignment of security measures with business objectives is a strategic need that demands a methodical assessment of its impact and value to the organization. This necessitates a set of well-defined metrics and **key performance indicators** (**KPIs**) that can provide tangible evidence of the effectiveness and benefits derived from such alignment. Furthermore, understanding the ROI and evaluating the holistic impact of these security measures is crucial for ensuring operational resilience, enhancing brand reputation, and garnering customer trust.

Metrics and KPIs

Establishing relevant metrics and KPIs is the initial step toward quantifying the effectiveness of cybersecurity initiatives in achieving business objectives. These metrics could range from **incident response** (**IR**) times, the number of detected and remediated vulnerabilities, to user awareness levels. However, it's important to go beyond technical metrics and develop KPIs that resonate with business outcomes. For instance, measuring the reduction in downtime, cost savings from avoided incidents, and improvement in regulatory compliance posture can provide a more tangible reflection of the business impact. Furthermore, customer-centric metrics such as the rate of successful phishing attempts can translate to customer trust and data protection. Let's take a closer look at this:

- **Example KPIs might include the following**:

  - **Mean time to detect** (**MTTD**) and **mean time to respond** (**MTTR**) to security incidents

  - % reduction in the number of security incidents

  - Compliance score with industry standards and regulatory requirements

  - Customer retention and acquisition rates post-implementation of enhanced security measures

  - Cost savings from cybersecurity initiatives are measured against the cost of potential breaches

- **Understanding the ROI**:

  Assessing ROI on aligned security initiatives is complex yet critical. Security leaders who successfully navigate through it often set up their organization for success as they can clearly outline why cybersecurity plays a critical role in the business. It entails a thorough analysis of financial implications, including cost savings from avoided security incidents, reduced legal and regulatory fines, and improved operational efficiency. An accurate ROI assessment helps justify security investments and provides a compelling narrative to the stakeholders on the value delivered through cybersecurity initiatives.

- **Operational resilience**:

  Aligned security measures can significantly contribute to operational resilience by ensuring the organization can withstand and recover from cyberattacks or data breaches. A resilient operation indicates a mature cybersecurity posture intricately tied to BC and **disaster recovery** (**DR**) planning. Evaluating the impact on operational resilience entails assessing the organization's capability to maintain critical functions during adverse events and its agility in restoring normal operations post-incident.

- **Brand reputation**:

  In a digital age, brand reputation is immensely influenced by an organization's cybersecurity posture. Customers, partners, and stakeholders often perceive the brand through its ability to protect sensitive data and ensure privacy. Ask yourself, if your primary bank was known to have a weak cybersecurity practice, would you continue working with them? A solid cybersecurity framework, aligned with business objectives, sends a strong positive signal to the market about the organization's seriousness in maintaining a secure and trustworthy environment.

- **Customer trust**:

  Customer trust often hinges on an organization's ability to safeguard sensitive data. Aligned security measures provide a structured approach to ensuring data privacy and security, enhancing customer trust. Clear communication on the organization's security posture and proactive measures in addressing cybersecurity concerns further solidify customer trust.

The holistic evaluation of the impact and value of aligned security initiatives is exhaustive. However, it's an indispensable effort that unveils the multidimensional benefits delivered through a cohesive cybersecurity strategy.

In conclusion, measuring the impact and value of aligned security initiatives is not merely a technical or financial assessment but a strategic endeavor reflecting the organization's commitment to achieving a harmonious balance between business aspirations and cybersecurity imperatives. Through a meticulous evaluation and continuous improvement approach, organizations are better positioned to realize the full potential of cybersecurity as a strategic business enabler, thereby steering the enterprise toward a future of resilience, growth, and sustained customer trust in an increasingly complex and threat-prone digital landscape. Now that the importance and interplay of risk assessment and **business impact analysis** (**BIA**) are established, let's delve deeper into the structured framework that refines this prioritization process. This will equip organizations with the necessary tools to effectively prioritize their security initiatives.

Prioritizing security initiatives based on risk and business impact

Let's delve into a structured approach toward making informed decisions on where to channel resources effectively to bolster cybersecurity posture while supporting business objectives. This approach's heart lies in a thorough risk assessment and BIA. By understanding the fundamentals of risk assessment, organizations can identify potential threats and vulnerabilities, laying the foundation for informed prioritization. A BIA can help to identify assets and processes critical for the organization's sustenance and growth. The interplay between these two assessments is crucial to deriving a priority list of security initiatives that resonate with the business objectives. Introduction to a structured framework further refines the prioritization process. This framework provides a scaffold for mapping real-world scenarios aligning security initiatives with business objectives. Organizations can choose an approach that best suits their unique operational landscape and threat profile by evaluating different frameworks and methodologies.

The significance of a well-crafted communication strategy is also critical. Engaging cross-functional teams and conveying prioritized security initiatives to stakeholders fosters a culture of collective cybersecurity consciousness and ensures the implementation is aligned with organizational goals.

The journey from assessing risks and business impact through prioritizing security initiatives to communicating and implementing them is a meticulous process requiring a blend of technical acumen and strategic foresight. We aim to equip you with the ability to prioritize security initiatives effectively and advocate for them in a manner that underscores their value to the business, thereby fostering a robust cybersecurity culture aligned with business growth and sustainability.

The importance of risk assessment and BIA

Risk assessment is a structured approach to identifying potential impacts, including vulnerabilities and threats to an organization. Risk assessments begin with identifying assets, then detecting vulnerabilities within those assets, and evaluating threats poised to exploit these vulnerabilities. A subsequent step involves quantifying the potential impact and likelihood of these threats materializing, which ultimately leads to a risk rating. This rating is instrumental in steering decision-making concerning allocating resources toward cybersecurity initiatives.

In cybersecurity, risk assessment is more than just a compliance checklist; it mirrors an organization's security posture. It helps to identify critical nodes where a security breach could result in substantial business disruption or financial loss. The BIA concept also steps into the spotlight. Conducting a BIA involves identifying and evaluating the potential effects of interruptions to critical business operations through various scenarios. It aids in determining essential functions and processes crucial for the organization's survival and assessing the potential operational and financial impacts of disruptions. These evaluations typically include lost sales and income, increased expenses, regulatory fines, and contractual penalties.

Furthermore, BIA aids in identifying tangible and intangible assets critical for maintaining BC. For instance, identifying data as a critical asset is the first step toward understanding the importance of securing it against unauthorized access or breaches. This information is invaluable as it sets the stage for prioritizing cybersecurity initiatives.

The crossroads of risk assessment and BIA is where the magic happens in sculpting a resilient cybersecurity posture. The interplay between risk assessment outcomes and BIA is pivotal in identifying priority areas for security investments. By understanding the risks facing the organization and the business impacts associated with those risks, a clear roadmap emerges, highlighting areas where security measures are critically needed.

For instance, if the risk assessment unveils a high likelihood of phishing attacks, and the BIA shows that a successful phishing attack could lead to significant financial loss and reputational damage, it's evident that anti-phishing training and technologies would be a priority. Hence, this analysis enables crafting a cybersecurity strategy that is robust and aligned with the organization's broader business objectives.

The discourse exemplifies the synergy between risk management, BIA, and the prioritization of security initiatives, which is crucial for cultivating a resilient business ecosystem. By carefully evaluating the risks and understanding the business impacts, organizations can architect a cybersecurity strategy that resonates with their business goals, propelling them on a sustainable growth trajectory with a fortified security posture.

Prioritizing security initiatives with frameworks

A well-thought-through and defined framework for prioritizing security initiatives based on assessed risks and business impacts is a compass in this venture. This framework is not initially linked to a specific methodology; instead, it should adopt a generic stance, paving the way for a tailored approach based on an organization's unique landscape.

The essence of this framework will hinge on a comprehensive understanding of the risks an organization faces and the cascading impact these risks could have on business objectives. Through a lens of risk assessment and BIA, organizations can sift through the noise and hone in on security initiatives that warrant priority. For instance, an e-commerce platform grappling with cyber threats of data breaches could leverage this framework to prioritize encryption, **multi-factor authentication** (**MFA**), and regular security audits. These measures fortify the security posture and align with business objectives of customer trust and regulatory compliance.

Many frameworks and methodologies, each with unique strengths and weaknesses, can be leveraged to operationalize this generic approach. One such framework is the **National Institute of Standards and Technology Risk Management Framework** (**NIST RMF**), which offers a structured process integrating security and risk management activities within the system development life cycle. Encompassing seven distinct steps ranging from preparation to continuous monitoring, RMF caters predominantly to government entities and substantial enterprises. RMF provides a holistic approach toward risk management, enabling a thorough comprehension and addressing security requisites in alignment with business objectives. Additionally, adherence to RMF can significantly aid in achieving compliance with myriad regulatory mandates and standards, thereby mitigating legal and compliance-associated risks. The emphasis RMF places on continuous monitoring is a cornerstone for the timely identification and alleviation of risks, rendering it a robust framework for organizations seeking a comprehensive methodology to intertwine security initiatives with business strategy. While RMF offers a comprehensive roadmap, its complexity might pose challenges for smaller entities or those with limited resources.

In contrast, the **Factor Analysis of Information Risk (FAIR)** model offers a quantitative lens through which risks can be analyzed and communicated, which is especially appealing when translating cybersecurity risks into financial implications. Distinguished for its quantitative risk analysis model, the FAIR methodology facilitates organizations in deciphering, analyzing, and quantifying information risk in financial terminologies. This framework finds its core utility within financial sectors or organizations endeavoring to articulate cyber risk in monetary terms, fostering a better alignment of security initiatives with business objectives. By translating abstract cyber risks into tangible financial terms, FAIR aids in elucidating the financial implications of risks to non-technical stakeholders. The quantitative insights gleaned from FAIR are instrumental in fostering informed decision-making concerning resource allocation and prioritization of security initiatives. This attribute of FAIR can be pivotal for organizations seeking a quantifiable risk analysis to guide strategic security investments and risk management endeavors.

Furthermore, the **Center for Internet Security (CIS)** provides a pragmatic checklist of actions to bolster cybersecurity posture. The CIS Controls embody a set of 20 pragmatic and actionable controls meticulously prioritized to enhance an organization's cyber defense posture. Tailored for a broad spectrum of industries, these controls are especially pertinent for small to medium-sized organizations questing for a practical and unambiguous approach to bolster cybersecurity posture. The CIS Controls simplify complex cybersecurity challenges into straightforward, actionable advisories, enabling a cost-effective methodology to tackle prevalent, high-impact cybersecurity issues. The community-driven ethos of the CIS Controls ensures a collective intelligence approach to cybersecurity, melding a myriad of insights from IT and cybersecurity professionals. This framework is a quintessential choice for organizations seeking a cost-effective, practical, and community-backed approach to improving cybersecurity while aligning with business objectives. However, it might overlook certain qualitative aspects, potentially needing more nuanced insights. While straightforward, it may need more analytical depth encapsulated in frameworks such as RMF.

Each of these frameworks, when wielded judiciously, can significantly enhance the alignment between security initiatives and business objectives. The choice of framework, or even a hybrid approach, would be dictated by the organization's industry, size, risk profile, and operations.

In assessing these frameworks, organizations embark on a journey of enhancing cybersecurity and intertwining it with the business fabric, ensuring that security initiatives are not mere reactionary measures but strategic enablers. Through a meticulous application of a suitable framework, organizations can architect a cybersecurity strategy that is both robust and business-centric, thus encapsulating the essence of aligning security initiatives with business objectives in a practical, actionable manner.

Communicating prioritized security initiatives

The delicate balance of information security with dynamic business objectives requires a well-structured communication strategy. This ensures that stakeholders, from the executive leadership to operational teams, are informed and in sync with the organization's cybersecurity vision.

To start, developing a robust communication strategy is essential. It serves as the core between the technicalities of security risks and broader organizational goals. For instance, instead of merely presenting that a particular system is vulnerable to specific **Common Vulnerabilities and Exposures (CVEs)**, a more comprehensive communication might focus on potential business disruptions or financial losses, drawing a more transparent line to business objectives. Effective communication isn't about overwhelming stakeholders with technical jargon. Instead, it's about clarity, relevance, and resonance. Consider a multinational organization looking to migrate its data to the cloud. Rather than diving deep into the intricacies of encryption protocols, the communication strategy could highlight the benefits of enhanced data accessibility and assurances of risk-mitigated data protection. This not only clarifies the initiative but also builds trust among stakeholders.

Beyond communication, the success of implementing these prioritized security initiatives lies heavily in cross-functional collaboration. Cybersecurity isn't a siloed function but an organizational imperative. Engaging cross-functional teams, from finance to operations to IT, fosters a collective commitment. For instance, introducing a new access control protocol isn't solely about the IT team ensuring its deployment. It would involve HR for employee training, finance for budget allocations, and operations to ensure minimal disruptions. Consider a company implementing MFA for its user accounts in a practical scenario. While the tech team integrates the MFA system, the marketing team might work on awareness campaigns, and customer support prepares for potential user queries. The synergy of these diverse teams ensures smoother and more effective implementation.

Lastly, gauging the effectiveness of these security initiatives is the final yet perpetual step. The metrics used to measure success should align with technical objectives and broader business goals. Implementing an advanced firewall system thwarts cyberattacks and ensures continuous service availability, leading to customer trust and satisfaction. For instance, after implementing a new transaction monitoring system, a financial institution should track reduced fraud incidents and observe customer response times, potential system downtimes, and customer feedback. The end goal is not just fortifying defenses but ensuring that, in doing so, the business thrives and remains resilient.

In conclusion, the journey of prioritizing and implementing security initiatives is comprehensive, iterative, and collaborative. It demands clarity in communication, integration across departments, and a constant review of its effectiveness in the ever-evolving cyber landscape. Organizations can bolster their defense mechanisms while fostering growth and resilience by rooting security initiatives in business objectives and ensuring consistent communication and collaboration.

Communicating the value of security investments

In this chapter, we will unveil the importance of communicating the value of security investments. Mastering this skill will be the pathway through which you, as a security professional, can articulate the business value derived from robust security postures, thus ensuring a holistic understanding and garnering unwavering support from stakeholders.

The necessity of translating technical metrics into business value becomes evident as we traverse the nuances of communication in the cybersecurity domain. The objective is to map the labyrinth of technical security metrics to tangible business outcomes, thereby forging a discernible link between security investments and business value. Through practical examples, this section elucidates how a well-secured environment can propel business objectives forward through risk mitigation, regulatory compliance, or customer trust.

Advancing further, the pivotal role of tailored communication strategies comes to the forefront. The section on developing effective communication strategies unveils the art of tailoring messages for non-technical stakeholders, ensuring clarity and resonance. Leveraging visuals and analogies, the narrative aims at simplifying complex security concepts, thus enabling a better understanding of the value proposition.

Moreover, engaging and building trust with stakeholders is an enduring process. Establishing regular communication channels and involving stakeholders in security investment decisions fosters a culture of transparency and collaborative decision-making. Through real-world examples, this chapter highlights how transparent communication can pave the way for enhanced stakeholder trust and support, which is paramount for the successful execution of security initiatives.

This chapter is crafted to equip security professionals with the strategies and insights necessary for effectively communicating the value of security investments, ensuring understanding, and fostering a collaborative environment conducive to upholding a robust security posture aligned with business objectives.

Translating technical metrics to business value

Technical metrics are critical; they clearly measure an organization's security posture. However, these metrics may be abstract to non-technical stakeholders who operate in the realm of business value and objectives. The bridge between these two disparate areas lies in translating technical metrics to a business value understandable to all stakeholders. This section dives into this integral process, aiding in aligning cybersecurity endeavors with business goals. In our mission to bridge the gap between technical metrics and business value, we need to follow a systematic approach. This approach requires a transition from understanding to action, which can be broken down into three key steps: understanding and identifying key business metrics, mapping technical security metrics to business metrics, and translating technical metrics to business value. Let's delve into each of these steps in detail:

- **Understanding and identifying key business metrics**:

  The first step is to grasp and pinpoint the critical business metrics that most matter to our stakeholders. It's essential to discern different stakeholders' varying concerns: for instance, a CFO might be keen on cost saving and ROI, while a COO might focus on operational efficiency and resilience. Standard business metrics include cost-effectiveness, operational efficiency, market reputation, customer satisfaction, and regulatory compliance. Understanding these metrics facilitates a common ground where technical and business narratives can converge.

- **Mapping technical security metrics to business metrics**:

  Once the critical business metrics are understood, the task is to map these to technical security metrics. This is a nuanced endeavor that requires a keen understanding of both cybersecurity and business domains. For instance, the technical metric of "time to detect and respond to incidents" can map to business metrics of operational efficiency and regulatory compliance. Similarly, "the percentage of systems patched" could relate to risk reduction and thus cost savings in potential incident handling.

  Understanding how technical security measures such as **intrusion detection systems (IDS)**, **endpoint detection and response** (EDR), firewalls, and encryption protocols contribute to BC, risk reduction, and compliance is vital. This mapping exercise is the core of communicating the value of security investments as it creates a clear line of sight from technical actions to business value.

- **Translating security investments to business value**: The following examples illustrate how strategic investments in cybersecurity can deliver tangible business value. From minimizing operational disruptions to fostering customer trust and achieving cost savings, a robust security infrastructure can influence key business metrics in significant ways. By translating these technical achievements into business outcomes, organizations can better communicate the importance and impact of their security investments:

  - **Example 1 – Reducing downtime**: Consider a retail company that invested in **distributed denial of service (DDoS)** prevention tools. Before the investment, the company experienced several hours of downtime due to DDoS attacks, leading to substantial revenue loss. The downtime was significantly reduced post-investment, improving operational efficiency and customer satisfaction. The technical metric of "fewer DDoS incidents" translates directly to business value by improving availability and protecting revenue streams.

  - **Example 2 – Improving customer trust**: A financial institution implementing robust encryption protocols to safeguard customer data can be another illuminating example. Here, the technical metric of "data encryption" translates to the business value of enhanced customer trust and regulatory compliance, which results in a higher customer retention rate and potentially lesser fines from regulatory bodies.

- **Example 3 – Cost savings**: Imagine an enterprise that invested in an advanced threat detection system, which resulted in fewer successful phishing attacks. The technical metric – "reduced phishing incidents" – translates to a business metric of cost saving, as fewer incidents mean lesser costs in incident remediation, potential legal liabilities, and regulatory fines.

These examples showcase a clear trajectory from technical metrics to quantifiable business value. They elucidate how investment in cybersecurity tools and protocols translates into tangible benefits, resonating well with the business objectives.

Translating technical metrics to business value is critical to aligning cybersecurity initiatives with business objectives. This exercise supports better communication with non-technical stakeholders and fosters a culture of shared responsibility toward achieving a robust security posture that propels business objectives forward. Through understanding, mapping, and demonstrating practical examples, the value of security investments becomes apparent and actionable to all stakeholders involved, paving the way for a more resilient and business-aligned cybersecurity strategy.

Developing effective communication strategies

Effective communication involves understanding the audience's baseline knowledge and delivering a resonant message. In cybersecurity, where technical jargon can quickly become a barrier, developing effective communication strategies is very important, especially when articulating the value of security investments to non-technical stakeholders. Let's explore the core aspects of crafting and delivering messages that bridge the technical-business divide, facilitating better comprehension and collaborative decision-making. To effectively communicate the value of cybersecurity investments, adopting a strategic approach that resonates with non-technical stakeholders is vital. This strategy can be broken down into the following key aspects:

- **Tailoring the message for non-technical stakeholders**:

  Always remember that the various stakeholders come from diverse functional backgrounds, each with a unique set of concerns and understanding. Tailoring the message to suit their comprehension levels and address their concerns is crucial. This involves avoiding technical jargon, focusing on benefits rather than features, and explaining the impact on business objectives rather than delving into technical specifications.

  For example, instead of detailing the technical workings of a new encryption protocol, explain how it helps safeguard sensitive customer data, thus enhancing trust and potentially increasing customer retention. This way, the message is tailored to resonate with non-technical stakeholders, addressing their primary concern of customer satisfaction and trust.

- **Leveraging visuals and analogies for better understanding**:

    Humans are visual creatures; a picture is often worth a thousand words. Leveraging illustrations such as infographics, graphs, or charts can help simplify complex concepts. For instance, showing a graph of incident reduction post-implementation of a security measure can visually represent the investment's value. Similarly, analogies can be powerful tools in explaining complex cybersecurity concepts. For example, comparing a firewall to a nightclub bouncer checking IDs before allowing entry can help stakeholders understand its function and value. Analogies help in translating technical terms into scenarios that are relatable to the audience, thereby aiding in better comprehension.

- **Case studies – effective communication of security investments**:

    Case studies provide a narrative explaining and demonstrating security investments' value. Real examples where security investments yield significant business benefits are compelling stories that engage stakeholders. Examples include the following:

    - **Example 1– Retail company's DDoS prevention investment**: Consider a retail company that invested in DDoS prevention tools and witnessed a significant reduction in downtime, thereby saving substantial revenue that would have been lost. Detailing this case, along with quantifiable data on revenue saved, can create a compelling narrative that underscores the value of the investment.

    - **Example 2 – Financial institution's encryption protocol implementation**: A case where a financial institution bolstered its customer trust and met regulatory compliance by investing in robust encryption protocols can serve as an exemplary story. Quantifying the increase in customer retention rates and the reduction in compliance-related fines post-implementation provides a tangible measure of the investment's value.

    - **Example 3 – Enterprise advanced threat detection system**: An enterprise that invested in an advanced threat detection system, consequently experiencing a drop in successful phishing attacks, can serve as a case in point. Detailing the cost saved in incident remediation and potential legal liabilities provides a clear picture of the ROI from this security investment.

Cybersecurity professionals can develop effective communication strategies by meticulously tailoring the message, leveraging visuals and analogies for better understanding, and presenting compelling case studies. These strategies enlighten non-technical stakeholders on the value of security investments and foster a culture of shared understanding and collaborative decision-making toward achieving common organizational objectives. In this light, effective communication transcends the boundary of merely conveying information; it becomes a conduit for engendering a security-conscious organizational culture aligned with business goals.

Engaging and building trust with stakeholders

Establishing a solid rapport between cybersecurity experts and the wider stakeholder group, including non-technical ones, heavily rests on engagement and trust. This connection is achieved through transparent communication, encouraging collective decision-making, and demonstrating the practical benefits of investing in security measures. There are several strategies one can implement to bolster trust with stakeholders:

- **Establishing regular communication channels**:

  Effective communication requires ongoing engagement and cannot be assumed to be achieved through a one-off task. Regular communication channels such as monthly security briefings, quarterly reviews, virtual open hours, or even casual coffee catch-ups can go a long way in keeping stakeholders informed about cybersecurity initiatives. For instance, a monthly security bulletin outlining key metrics, recent incidents, and upcoming initiatives could be a straightforward, concise communication tool. Furthermore, leveraging digital platforms such as intranet forums or chat groups can foster continuous engagement and provide stakeholders with a platform to raise concerns or seek clarifications.

- **Building stakeholder trust and support through transparent communication**:

  Transparency in communication is a cornerstone in building trust. Leading by transparency is paramount for it. Sharing successes and challenges in security initiatives fosters a culture of honesty and continuous improvement:

  - **Example 1**: A company could share a case where an investment in a robust data encryption solution prevented a potentially damaging data breach, thus preserving patient trust and avoiding regulatory fines.

  - **Example 2**: Similarly, sharing benchmarking studies showcasing how the organization's security posture compares with industry standards or peers can also enlighten and reassure stakeholders about the effectiveness of current security investments.

- **Collaborative decision-making – involving stakeholders in security investment choices**:

  Promoting a sense of ownership among your stakeholders helps ensure that security investments are aligned with broader business objectives. For example, you can organize regular sessions where stakeholders can voice their concerns, provide input, or challenge proposed security investments worth considering. For instance, a collaborative workshop to prioritize security initiatives for the upcoming fiscal year can promote an environment of transparency and inclusivity. Through such collaborative efforts, security professionals can ensure that proposed security investments resonate with stakeholder expectations and business objectives.

Always remember that engaging and building trust with your stakeholders isn't a checkbox task but a long-term commitment. To be successful, you will require a balanced approach to regular communication, collaborative decision-making, and transparent disclosure. The strategies discussed provide a roadmap for all security professionals to foster meaningful relationships with stakeholders, ensuring that the cybersecurity narrative is understood and valued across the organizational spectrum. Organizations can create a conducive environment for cybersecurity initiatives to thrive through such engagements, ultimately supporting and advancing overarching business objectives.

Summary

In this chapter, we explored the importance and methodology of translating technical cybersecurity metrics into business values. We emphasized the criticality of effective communication, particularly when conveying the value of cybersecurity investments to non-technical stakeholders. The strategies discussed included tailoring messages for varied stakeholders, leveraging visuals and analogies, and using case studies for better understanding. We also emphasized the role of regular and transparent communication in building trust and facilitating collaborative decision-making. The chapter ended by underscoring the significance of stakeholder engagement and the long-term commitment required to build meaningful relationships. In sum, the chapter provided a comprehensive guide on aligning cybersecurity initiatives with business objectives and fostering a culture that values and understands the importance of cybersecurity. Looking ahead, you will delve deeply into how to discern truth from marketing claims as you work with security vendors.

7
Demystifying Technology and Vendor Claims

As cybersecurity professionals, we are bombarded daily with new acronyms, sales pitches, and products that promise unmatched protection and innovation. However, the reality is often more complex than these claims suggest, so it's essential not to trust but always to verify. To evaluate these technologies, we must first understand our organization's needs and assess their potential benefits carefully.

Vendors often tout their solutions as the answer to all cybersecurity challenges. However, discerning the truth in these claims requires knowledge, experience, and a critical approach. Understanding these claims entails technical expertise, asking the right questions, and seeking supporting evidence.

In this chapter, we'll help you understand the often complex world of vendor claims by covering the following topics:

- **Understanding technology and vendor claims**: This involves recognizing the need to critically evaluate what vendors say about their products and technologies. It's about not taking claims at face value and understanding the importance of scrutiny.

- **Critically analyzing claims**: This highlights the process of questioning and dissecting vendor statements. It involves applying skepticism to assess the truthfulness and practicality of these claims, which is essential in an industry filled with both innovative solutions and exaggerated marketing.

- **Utilizing analyst and third-party testing reports**: This emphasizes the significance of leveraging external resources such as reports from analysts and third-party testing organizations. These reports serve as invaluable tools to either confirm or refute vendor claims, providing an objective basis for evaluation.

- **Thoroughly assessing vendors**: This focuses on the necessity of comprehensive evaluations that go beyond mere technical specifications. It includes examining the vendor's reputation, financial health, and dedication to customer support. Such assessments ensure that the selected solution aligns with both the technical needs and the broader objectives and values of the organization.

By the end of this chapter, you will have the necessary skills to effectively recognize, analyze, and evaluate technology and vendor claims.

Understanding technology and vendor claims

This section is designed to guide you through this intricate landscape, offering clarity and insight. As cybersecurity professionals, our decision-making process is often influenced by the claims and promises made by technology vendors. Always remember: don't fall for marketing lingo and claims, but verify that the technology fits your needs. The cybersecurity industry is filled with specialized terminologies and metrics; understanding these is crucial. By familiarizing yourself with this language, you'll be better equipped to interpret what vendors offer.

But understanding the language is just the beginning. We also need to develop the skill of separating fact from marketing. Vendors often use persuasive language and bold assertions to stand out in a competitive market. Identifying what is a genuine capability and what is mere hyperbole is critical to making informed decisions.

In this section, we'll delve into the substance behind the claims. Evaluating what is being said and the practical implications of these claims is essential. How do these technologies work in real-world scenarios? Are they as effective as claimed? These are some of the questions we will address.

Deciphering the language of cybersecurity claims

Cybersecurity is marked by rapid technological advancements and evolving threat landscapes, leading vendors to innovate and update their offerings constantly. In this environment, terms such as *machine learning*, *AI-driven security*, *real-time monitoring*, and *automated threat detection* frequently surface in vendor pitches. While these terms promise advanced capabilities, their real-world application often varies, necessitating a more profound understanding beyond the surface-level allure. This understanding goes beyond knowing what these terms mean; it involves asking the right questions to assess the solutions' capabilities and limitations.

Machine learning and *AI-driven security* suggest a high degree of automation and intelligence in cybersecurity solutions. However, their actual application is usually more assistive, augmenting human analysts in identifying patterns and anomalies. These technologies are crucial in sifting through vast amounts of data to detect potential threats, yet they only partially replace the nuanced judgments made by experienced cybersecurity professionals.

Questions to ask your vendor:

- What is the role of human oversight in the AI decision-making process, and how does this affect response times to emerging threats?

- How is machine learning integrated into your solution, and what threats is it most effective against?

- Can you provide examples of how AI has successfully identified and mitigated threats in real-world scenarios?

Similarly, *real-time monitoring* and *automated threat detection* imply a proactive stance in cybersecurity defense. Real-time monitoring refers to continuously surveilling activities, a critical component in swiftly identifying and mitigating potential threats. Automated threat detection, on the other hand, suggests a system's capability to recognize and address security threats independently. However, the extent of this automation can differ significantly among products, often requiring a combination of technology and human oversight for optimal performance.

Questions to ask your vendor:

- **Real-time monitoring**:

  - What does real-time monitoring entail in your solution, and how quickly can it detect and alert potential threats?

  - What metrics do you use to track your detection efficiency?

  - How do you define real-time monitoring? For example, is up to five minutes to detect issues considered real time for you?

  - How does the system differentiate between false positives and true positive threats in real-time monitoring?

  - Can you demonstrate how real-time monitoring has helped other organizations in similar sectors or with similar threat profiles?

- **Automated threat detection**:

  - What types of threats does your automated threat detection system primarily target, and how does it learn and adapt to new threats?

  - How do you ensure the automated detection system stays up-to-date with the latest threat intelligence?

  - What mechanisms exist for human intervention in cases where automated detection is insufficient?

Other key terms such as *end-to-end encryption*, *compliance management*, and *threat intelligence* are equally important in the cybersecurity dialogue. End-to-end encryption is essential for securing data during transmission and safeguarding it from unauthorized access. Compliance management points to the adherence to legal and regulatory standards in cybersecurity practices, a non-negotiable aspect in today's heavily regulated digital world. Threat intelligence involves collecting and analyzing information about existing or emerging threats, which is vital in developing robust cybersecurity strategies.

Questions to ask your vendor:

- **End-to-end encryption**:

  - How do you implement end-to-end encryption, and what standards or protocols are used?

  - Are there any scenarios where the data is decrypted within your system, and if so, how is this data secured?

  - How does your end-to-end encryption handle key management and ensure that only intended recipients can decrypt the data?

- **Compliance management**:

  - How does your solution help ensure compliance with specific regulations relevant to our industry, such as GDPR or HIPAA?

  - Can your system generate compliance reports, and how frequently can these be updated or customized?

  - How does your solution adapt to changes in compliance requirements or standards?

- **Threat intelligence**:

  - What threat intelligence sources do you utilize, and how is this information integrated into your solution?

  - How frequently is your threat intelligence updated, and what processes are in place to ensure its relevance and accuracy?

  - Can you provide examples of how your threat intelligence has preemptively helped mitigate potential security breaches?

Using these terms in vendor communications can sometimes be more marketing-driven than reflective of actual capabilities. It is not uncommon for vendors to use these buzzwords to create an aura of sophistication around their products, which may not fully align with their practical functionality. This gap between marketing claims and real-world applications underscores the need for cybersecurity professionals to understand these terms keenly.

In learning to decipher the language of cybersecurity claims, professionals equip themselves with the ability to evaluate vendor communications critically. This skill is indispensable in assessing the suitability and effectiveness of different cybersecurity products and services. It enables professionals to ask the right questions, seek clarifications, and make decisions based on an accurate understanding of what these solutions can genuinely offer.

In conclusion, the language of cybersecurity is not just a collection of buzzwords but a key element in the dialogue between vendors and professionals. Understanding this language is crucial for interpreting communications accurately and making informed decisions. As the cybersecurity landscape continues to evolve, so will its language, and staying abreast of these changes is essential for any professional committed to safeguarding their digital environment.

Separating facts from marketing in vendor claims

The ability to discern fact from fiction in vendor claims is more than just a skill—it's a necessity. Let's delve into the critical challenge of sifting through the often persuasive and sophisticated marketing language vendors use to uncover the actual capabilities and limitations of their cybersecurity solutions.

Vendors may use compelling language and bigger-than-reality claims about their products to capture market share. These claims, while not necessarily false, can often be exaggerated or presented so that they create unrealistic expectations. The primary challenge for cybersecurity professionals is to navigate these claims critically, separating the wheat from the chaff.

The skill of distinguishing between factual statements and marketing hyperbole is not innate; it is developed through experience, knowledge, and a keen understanding of the cybersecurity landscape. This process involves scrutinizing the presented information, asking probing questions, and seeking evidence to support or refute the claims.

The following are a few techniques for analytical evaluation:

- **Questioning the evidence**: One of the first steps in evaluating vendor claims is to ask for evidence that supports their assertions. This could include independent test results, case studies, customer testimonials, or performance benchmarks.

- **Understanding the context**: It's crucial to understand the context in which these claims are made. For instance, a vendor might claim their product offers *unparalleled security*, but understanding the specific security needs of your organization is necessary for this claim to be clear and substantiated.

- **Seeking independent verification**: Independent reviews, user feedback, and industry analysis can provide an unbiased perspective on a product's performance and reliability. This external validation can be instrumental in verifying the vendor's claims.

- **Recognizing common marketing tactics**: Awareness of common marketing tactics, such as fear-based selling, where vendors play on the fears of cybersecurity threats, can help professionals remain objective. Similarly, understanding that terms such as *market-leading* and *cutting-edge* are subjective and can be misleading is essential.

Developing a critical mindset is vital in this process. As cybersecurity professionals, we must approach vendor claims with healthy skepticism, not dismissing them outright but rigorously evaluating their veracity. This involves being aware of the psychological impact of marketing language and consciously focusing on the factual content. Simply speaking, always verify.

The cybersecurity field is continuously evolving, and so are marketing strategies. Keeping abreast of the latest trends, technologies, and threats is essential. This ongoing education aids in better understanding and evaluating new products and services as they emerge in the market.

The ability to separate facts from marketing hyperbole in vendor claims is essential for cybersecurity professionals. This chapter equips readers with the tools and techniques to critically analyze vendor communications, ensuring that decisions are made based on an accurate understanding of a product's capabilities and limitations. As professionals in this field, adopting a vigilant, inquisitive, and analytical approach is vital in navigating the complex and often misleading world of cybersecurity marketing. This skill not only aids in making informed decisions but also in safeguarding the digital assets and integrity of the organizations we protect.

Evaluating the substance of cybersecurity solutions

Let's pivot from simply understanding the language of vendor claims to critically assessing their real-world applicability and effectiveness. This evaluation is a crucial step in the process, as it determines how well these claims translate into tangible, effective cybersecurity solutions that can genuinely meet the specific security needs of an organization.

In cybersecurity, where theoretical claims often clash with practical realities, the ability to discern the actual substance of a solution is invaluable. In their efforts to market their products, vendors may present claims that, while technically accurate, might not fully align with an organization's practical needs or constraints. This discrepancy underscores the importance of evaluating what is promised and how these promises hold up in actual usage scenarios.

The following are a few techniques for effective evaluation:

- **Scenario-based testing**: One of the most effective ways to assess the practicality of a solution is through scenario-based testing. This involves simulating real-world situations or threats to see how the solution performs under conditions that mimic actual operational environments. Remember: trust, but verify. Ensure you test solutions before purchasing.

- **Compatibility and integration**: Another critical aspect is evaluating how well a solution integrates with existing systems and infrastructure. Compatibility issues can render even the most robust solutions ineffective if they cannot be seamlessly incorporated into the technological ecosystem.

- **Scalability and flexibility**: Assessing the scalability and flexibility of a solution is also crucial. The cybersecurity landscape is dynamic, and solutions must adapt to evolving threats and expanding business needs.

- **Cost-benefit analysis**: A thorough cost-benefit analysis is essential. This involves weighing the benefits of a solution against its costs, not just in terms of financial investment but also in terms of resource allocation, maintenance, and potential disruption during implementation.

In addition to these techniques, seeking external validation from independent reviews, user testimonials, and industry benchmarks can provide an unbiased perspective on the effectiveness of a solution. This external input can be beneficial in verifying vendor claims and understanding the solution's performance in diverse settings.

Post-implementation, continuous monitoring of the solution's performance is vital. This ongoing evaluation helps ensure the solution effectively meets organizational needs and adapts to changing security landscapes.

Before we move to the next topic, it's important to highlight the significance of effectively evaluating the claims made by vendors. This step takes us beyond understanding and allows us to practically assess how these claims hold up in real-world situations. It ensures that the cybersecurity solutions we choose are not just theoretically sound but truly effective in practice. By utilizing techniques such as testing based on scenarios, checking compatibility, assessing scalability, and analyzing cost-benefit factors, we position ourselves to make decisions that align with our organization's specific security needs and limitations. This process is strengthened by validation and continuous monitoring after implementation, which plays a role in selecting solutions that genuinely safeguard against evolving cyber threats. As we delve deeper into aspects of selecting cybersecurity solutions, let's remember that our goal is to find a fit that's both functional and seamlessly integrated into our strategic approach.

Critically analyzing claims

The world of cybersecurity is filled with complex solutions and sophisticated technologies, often presented with compelling claims by vendors. However, it's crucial to remember that these claims, while sometimes based in fact, can also be embellished with marketing flair. Understanding how to peel back the layers of these claims to reveal their actual substance is a skill every cybersecurity professional needs to develop.

This segment aims to equip you with a skeptical mindset—a critical tool in your arsenal. It's essential to learn to accept only some claims at face value, but rejecting them outright with proper evaluation is equally vital. Striking a balance between skepticism and an open mind is crucial in making well-informed decisions.

We will explore how to place vendor claims within the proper context. Every claim, every statistic, and every promised outcome exists within a specific framework—be it the particular needs of your organization, the broader cybersecurity landscape, or the vendor's marketing strategy. Understanding this context is crucial for assessing how relevant and applicable these claims are to your situation.

Furthermore, we will tackle the challenge of identifying biases and unsupported assertions in vendor communications. Like any other business, vendors have objectives and may present information to serve their interests. Recognizing these biases and asking for evidence to support claims is critical in navigating the sea of information.

By the end of this section, you will be equipped with the skills to hear, truly listen to, and critically evaluate what vendors are saying. This will enable you to make choices best suited to your organization's cybersecurity needs. Remember, in cybersecurity, a well-informed decision is your most vigorous defense. Welcome to a journey of developing a critical, analytical mindset that will empower you to sift through the noise and find the information that truly matters.

Developing a skeptical mindset

Adopting a skeptical mindset doesn't mean dismissing every claim outright; instead, it's about developing an analytical approach to evaluating information. This mindset is not just a defensive mechanism against misleading claims but also a proactive tool that aids in uncovering the truth. The art of skepticism in cybersecurity is about striking a balance—being cautious without being closed off, questioning without rejecting viable information, and discerning fact from marketing embellishment.

The first step in developing this mindset is learning to question everything. In a field where vendor claims range from nuanced truths to outright hyperbole, asking the right questions is critical. This means taking information at face value and probing deeper: What evidence supports this claim? How does this solution perform in real-world scenarios? Are there any independent reviews or benchmarks available? These questions help peel back the layers of marketing language to reveal the product's value.

It's crucial to distinguish between healthy skepticism and cynicism. While skepticism is a critical evaluation of claims based on evidence and logic, cynicism often leads to outright disbelief and distrust. Cynicism can close the door to potentially beneficial solutions, whereas skepticism keeps the door open but invites claims to be substantiated.

An essential aspect of skepticism is keeping an open mind to credible information. Dismissing a claim without consideration can lead to missed opportunities or overlooking valuable solutions. It involves assessing the source's credibility, the argument's logic, and the evidence provided. When a claim passes these tests, it's essential to be open to accepting it, even if it challenges preconceived notions or previous knowledge.

To maintain a healthy level of skepticism, continuous learning and staying updated with the latest trends and technologies in cybersecurity are crucial. Engaging with a community of peers for insights and experiences, attending industry conferences, and participating in professional forums can provide a broader perspective and aid in sharpening critical thinking skills.

Developing and maintaining a skeptical mindset is a critical skill. It empowers us to navigate vendor claims effectively, discerning the genuine from the exaggerated. It's essential to have the tools to cultivate this mindset and emphasize the importance of balancing skepticism with open-mindedness, ensuring that decisions are made based on a thorough and fair evaluation of available information. By mastering this skill, cybersecurity professionals can confidently select solutions that meet organizational needs, ensuring robust and effective cybersecurity defenses.

Contextual analysis of vendor claims

Let's focus on unpacking the various layers that form the backdrop against these claims. This analysis is crucial for evaluating the claims' relevance and validity and aligning them with your organization's needs and realities.

One must first grasp the broader industry landscape to analyze vendor claims effectively. This includes trends, emerging threats, technological advancements, and common challenges faced in the cybersecurity domain. A claim might seem impressive in isolation, but when viewed against the backdrop of industry standards and practices, its actual value (or lack thereof) can become apparent. For instance, a vendor might tout their solution as *revolutionary*, but its impact might be less significant than implied if it merely mirrors what is already available.

Vendors tailor their claims to resonate with their target audience, which can vary widely regarding technical expertise, industry focus, and specific security challenges. By understanding who the vendor is targeting, cybersecurity professionals can better assess the relevance of these claims to their organizations. A product that's ideal for a large financial institution with a robust IT infrastructure may be less suitable for a small nonprofit with limited technical resources, despite similar claims of effectiveness and security.

The market environment in which these claims are made also plays a critical role in their analysis. This includes competitive pressures, regulatory landscapes, and the economic climate. For example, vendors might make aggressive claims to stand out in a highly competitive market. Similarly, vendors may emphasize compliance aspects of their products in industries subject to stringent regulatory requirements. Understanding these dynamics helps in interpreting claims more accurately.

Perhaps most importantly, contextual analysis involves aligning vendor claims with your organization's needs. This includes considering factors such as the existing security infrastructure, the nature of the data being protected, employee skill levels, and budgetary constraints. A solution that excels in a particular area but does not align with your organization's unique requirements or constraints might not be the optimal choice, regardless of the vendor's claims.

To effectively analyze the context of vendor claims, staying informed about the latest cybersecurity developments is essential. Regularly engaging with industry news, reports, and peer discussions provides a solid foundation for this understanding. Maintaining a clear picture of your organization's cybersecurity posture and needs allows for a more targeted and practical assessment of vendor claims. To achieve that you can follow these steps:

Step 1: Identify the claim

First, identify the specific claim being made. This could be about the product's features, performance, or benefits compared to competitors.

Step 2: Gather information on the product and its context

Research the product thoroughly by gathering data from various sources such as the vendor's website, press releases, user reviews, industry reports, and expert opinions. This will help you understand the technology better and assess the credibility of the claim.

Step 3: Assess the validity of the claim

Compare the information gathered in step 2 with relevant data from other sources or similar products to determine whether the vendor's claim is accurate or exaggerated. Consider factors such as product specifications, performance tests, and user experiences.

Step 4: Evaluate the context

Consider the context of the claim within the industry and market trends. Is this a new innovation or an improvement on existing technology? Are there any competing products with similar claims? Understanding the broader context will help you assess whether the vendor's claim is significant or just a marketing strategy.

Step 5: Consider potential biases

Be aware of possible biases in the information provided by the vendor, such as self-promotion or exaggeration for marketing purposes. Cross-checking with independent sources can help minimize these biases and provide a more accurate assessment of the claim.

Step 6: Analyze the impact on your decision-making process

Consider how this claim might affect your decisions related to technology adoption, investment, or partnerships. If the vendor's claim is credible and significant, it could influence your choices in favor of their product. However, if the claim appears exaggerated or unsupported by evidence, you may want to reconsider its importance in your decision-making process.

Step 7: Share your findings with others

Share your analysis with colleagues, industry peers, and other stakeholders who might be interested in this technology vendor's claims. This will help validate your conclusions and provide a more comprehensive understanding of the claim within the broader context.

In summary, contextual analysis of vendor claims is a multifaceted process that requires understanding the broader industry landscape, the target audience, market dynamics, and, most importantly, your organization's specific needs and realities. By mastering this skill, cybersecurity professionals can ensure that their solutions are theoretically sound, practical, and beneficial to their organizations, enhancing their overall cybersecurity stance.

Identifying biases and unsupported assertions

Critical to evaluating cybersecurity vendor claims is identifying biases and unsupported assertions, and recognizing and challenging the biases and statements that can skew our understanding and decision-making. Discerning these elements is critical to making informed, objective vendor marketing and product evaluation decisions.

Vendors, driven by the objective of selling their products, can consciously or unconsciously introduce biases into their claims. These biases manifest as overly optimistic portrayals of a product's capabilities, downplaying potential limitations, or comparisons with competitors that put their products in an unfairly favorable light. For instance, a vendor might emphasize the strengths of their solution in ideal conditions without mentioning its performance under more typical or stressful scenarios.

As readers and evaluators of these claims, we also bring our own sets of biases. These could stem from previous experiences, brand loyalties, industry perceptions, or even the influence of peers and market trends. Such biases can cloud our judgment, making us favor one product without a thorough and objective evaluation. It's essential to be self-aware and question our assumptions and predispositions regularly.

Another critical skill is identifying unsupported assertions. These are claims that need more substantial evidence or rely on vague, ambiguous language. They can be as subtle as using terms such as *best in class* or *cutting edge* without concrete metrics to back them up, or as overt as making sweeping statements about a product's capabilities without any data or customer testimonials to support them.

The following are a few strategies for challenging biases and assertions:

- **Seeking evidence**: Always ask for evidence to back up claims. Having tangible proof is essential, whether it's performance data, independent reviews, or case studies.

- **Asking clarifying questions**: When faced with vague language or broad statements, ask specific, clarifying questions. This can help reveal the substance (or lack thereof) behind the claims.

- **Comparing with independent sources**: Cross-reference vendor claims with information from independent sources. This can include industry reports, user reviews, or expert opinions.

- **Reflecting on personal biases**: Take time to reflect on your biases and how they might influence your evaluation of the claims. Seeking a second opinion can also provide a fresh perspective.

In conclusion, identifying and challenging biases and unsupported assertions in vendor communications is an invaluable skill in cybersecurity. It allows professionals to see beyond the surface of marketing strategies and evaluate products based on their merits and fit for their specific organizational needs. As cybersecurity professionals, we must make more balanced, informed decisions that align with our objectives and requirements, ultimately strengthening our cybersecurity posture.

Making informed decisions on solutions is not just about understanding what vendors claim but also about seeking and interpreting objective external insights. This section is designed to effectively guide you through leveraging these invaluable external resources. We will focus on how to use reports such as the Gartner Magic Quadrant, the Forrester Wave, and MITRE ATT&CK Evaluations to validate vendor claims and ensure that solutions align with your organizational needs.

Utilizing analyst and third-party testing reports

The task of navigating through numerous cybersecurity solutions can be overwhelming. With each vendor presenting their product as the top choice, it becomes critical to seek unbiased external opinions. This is where renowned industry reports and evaluations come into play. For example, the Gartner Magic Quadrant provides a comprehensive analysis and comparison of vendors in a specific market, helping you understand who the key players are and how they perform against each other in various critical areas.

However, accessing these reports is just the beginning. A crucial aspect is interpreting them correctly. The Forrester Wave, for instance, offers detailed evaluations of products in a specific category, rating them on various criteria. Understanding how to read these reports, the criteria, and how they apply to your context is essential for deriving meaningful insights.

Interpreting these resources necessitates a critical and analytical approach. It's essential to recognize that while these reports are valuable, their relevance can vary based on your specific needs and context. Take the MITRE ATT&CK Evaluations, for example. These assessments focus on how effectively different cybersecurity products can detect and respond to simulated attacks based on real-world tactics. Understanding the nuances of these evaluations can help you gauge how a particular solution might perform in your IT environment.

Be cautious that vendors often find creative ways to declare victory or leadership, even when these third-party reports do not provide a clear benchmark or are not entirely favorable to them. This phenomenon is known as *marketing spin*, where vendors selectively highlight certain aspects of a report while downplaying or ignoring others that might not be as favorable. For instance, a vendor might be categorized as a *Leader* in the Gartner Magic Quadrant but may need to score better in certain critical capabilities that are crucial to your specific needs. Similarly, a vendor might showcase high performance in one or two criteria in the Forrester Wave but may still need to meet the essential requirements for your organization. In the case of MITRE ATT&CK Evaluations, vendors might emphasize their success in detecting specific attack scenarios while glossing over areas where their performance was suboptimal.

Applying the insights from these reports and evaluations to your organization's unique context is the final, critical step. This involves aligning the external information with your organization's cybersecurity challenges and goals. It's about making informed decisions that are not only based on robust external validation but are also customized to meet your organization's unique requirements.

Understanding and accessing external resources with practical examples

Let's explore the critical task of leveraging external sources for validating cybersecurity vendor claims. This section provides practical guidance on identifying, accessing, and effectively using various external resources such as analyst reports and third-party testing results.

In the cybersecurity field, several essential external resources offer valuable insights. For instance, the Gartner Magic Quadrant report can be accessed through Gartner's website, often requiring a subscription or registration. These reports categorize vendors into leaders, challengers, visionaries, and niche players, providing a comprehensive market overview. Similarly, Forrester Wave reports, accessible through Forrester's website or sometimes directly from vendors who have been evaluated, offer detailed analyses of vendors' offerings and market presence.

To access these reports, professionals can start by visiting the respective websites of Gartner and Forrester. Many organizations already have subscriptions to these services, so check with your IT or research department. If not, consider requesting access or looking for summaries and excerpts of these reports that are often available on vendor websites.

For third-party testing results, one valuable resource is the MITRE ATT&CK Evaluations. These can be accessed directly from the MITRE website, providing information on how different cybersecurity solutions perform against known attack techniques.

Navigating these reports effectively requires a focused approach. For instance, when using the Gartner Magic Quadrant, one practical strategy is identifying which quadrant aligns with your organization's needs. If you prioritize innovation, look at the *Visionaries* quadrant. For proven solutions, focus on *Leaders*.

When reviewing results such as those from MITRE, a practical approach is to focus on specific attack techniques most relevant to your organization. For example, if your organization is concerned about ransomware, consider how well solutions detect and respond to these specific attack scenarios.

To apply these insights to your organization, conduct a needs assessment first. Identify the critical threats you face, your existing cybersecurity infrastructure, and your budget. Then, cross-reference these with the strengths and weaknesses of vendors as highlighted in the analyst reports. For instance, if your assessment reveals a high risk of **advanced persistent threats** (**APTs**), focus on how vendors perform in this area as per the MITRE evaluations.

An important note in utilizing external resources is prioritizing direct information from third-party sources over vendors' interpretations or summaries. While many vendors provide summaries or highlights of analyst reports and testing results, there is a potential for these to be skewed or selectively presented to favor their products. Some vendors may twist words or cherry-pick data to cast their solutions in the best light, a common *marketing spin*.

Understanding and accessing external resources such as analyst reports and third-party testing results is a critical skill for cybersecurity professionals. These resources provide an objective perspective that is essential for validating vendor claims. By following the practical steps outlined in this section, such as accessing reports through official channels, focusing on relevant parts of these reports, and aligning their insights with organizational needs, professionals can make well-informed decisions in selecting the most suitable cybersecurity solutions.

Interpreting methodologies and results

This section clarifies the various testing approaches and methodologies used in external reports and tests. It will provide an in-depth understanding of how these methodologies can impact the conclusions drawn from these resources. Unraveling the complexities of these methodologies will help you with the straightforward interpretation of the results obtained from these resources.

Third-party tests can vary in their focus and methodology when evaluating cybersecurity solutions. While some tests may assess the solution's effectiveness in detecting and mitigating threats, others might concentrate on factors such as ease of use, implementation, and system performance impact. It is essential to understand each test's specific focus and methodology to draw meaningful insights.

Another essential skill is evaluating the rigor and relevance of the testing methodology. Rigor refers to the comprehensiveness of the testing or evaluation process, including a wide range of threat scenarios, from common malware to sophisticated attacks. Relevance pertains to how well the testing scenarios and evaluation criteria align with real-world cybersecurity challenges. A relevant test would simulate realistic operational environments and threat landscapes, providing insights that apply to actual cybersecurity scenarios. Mastering these skills ensures that cybersecurity solutions are adequately tested and evaluated for real-world effectiveness.

The accurate analysis of test results and reports is essential to make informed decisions. For example, when evaluating a cybersecurity solution that has received a high rating for its detection capabilities, it's essential to examine the conditions under which the rating was achieved. Were the tests conducted in a controlled environment by the vendor or under real-world conditions? How recent are the tests, and do they incorporate the latest threat intelligence?

A practical example of this skill in action can be observed in interpreting the results of the MITRE ATT&CK Evaluations. These evaluations test cybersecurity solutions against known attacker tactics and techniques. A comprehensive interpretation of these results would involve examining the overall scores, comprehending how each solution performed against specific tactics and techniques, and determining whether they align with the threats most pertinent to your organization.

When analyzing analyst reports such as the Gartner Magic Quadrant report or the Forrester Wave, it's essential to understand the criteria used for evaluation and how they relate to your organization's specific needs. A solution might be a leader in the quadrant, but the criteria for this ranking, such as user-friendliness, cost, or specific security features, may not correspond to your priorities.

It's more than just relying on internal evaluations and assessments. It's equally important to analyze external reports and tests. This skill enables you to make informed decisions based on a thorough and accurate understanding of how potential solutions will perform in real-world scenarios. By developing this skill, professionals can confidently select solutions rated highly in external evaluations that are tailored to their organizations' unique security requirements and challenges.

Applying findings to an organizational context

Translating the insights from external analyst reports and third-party tests into actionable decisions that suit an organization's requirements and context is essential. This section underscores the significance of comprehending external findings and utilizing them to align with your organization's distinct features, difficulties, and objectives.

The cybersecurity requirements of every organization are unique and depend on various factors, such as the nature of its business, the sensitivity of the data it handles, its regulatory environment, and its risk profile. The first step to applying external findings effectively is to assess how well they align with these specific needs. For example, if an organization deals with sensitive customer data, the focus might be on implementing solutions that offer robust data encryption and intrusion prevention capabilities. In this context, reports such as the Gartner Magic Quadrant or the Forrester Wave must be evaluated to determine their relevance.

Organizational constraints such as budget, existing IT infrastructure, and staff expertise significantly influence the selection of cybersecurity solutions. A highly-rated solution that requires a substantial investment in new infrastructure or extensive staff training might only be feasible for some organizations. Similarly, a solution that excels in large-scale enterprises may not be suitable for small- to medium-sized businesses. Therefore, interpreting the findings from external reports must include an analysis of these constraints.

Once the external findings have been assessed for applicability and fit within organizational constraints, the next step is to integrate these insights into the decision-making process. This involves a few key steps:

- **Prioritizing cybersecurity goals**: Based on the organization's risk assessment, prioritize what the cybersecurity solution needs to achieve—protecting against specific threats, compliance with regulatory standards, or ensuring minimal downtime.

- **Matching solutions to goals**: Use the insights from the external findings to identify which solutions best match these goals. For example, if mitigating insider threats is a priority, focus on solutions rated highly for user behavior analytics and access controls in third-party tests.

- **Engaging stakeholders**: It's essential to involve key stakeholders, including IT staff, management, and possibly even end users. Their input can provide additional perspectives on how well a solution fits the organizational context.

- **Pilot testing**: Conducting a pilot test of the shortlisted solutions can provide a practical understanding of how they would work in the organization's specific environment.

For example, organizations in the healthcare sector are increasingly concerned with cybersecurity solutions. The essential requirements are compliance with healthcare regulations, safeguarding patient data, and integration with existing health IT systems. To evaluate the most appropriate solutions, the organization's decision-makers will examine external findings demonstrating a solution's excellence in these areas. Additionally, budget and ease of integration with existing systems will be considered.

A pilot test can provide practical insights into the effectiveness of a shortlisted vendor's healthcare cybersecurity solution in the real-world context of the organization.

The skills to effectively apply the findings from external cybersecurity reports and tests to the unique context of your organization are essential. This application involves carefully assessing how these findings align with specific cybersecurity needs, considering organizational constraints and goals, and integrating these insights into a well-informed decision-making process. By mastering this skill, professionals can ensure that the cybersecurity solutions they select are highly rated in industry reports and the most appropriate and effective for their specific organizational contexts.

By considering and incorporating information from reports and independent tests, organizations can make well-informed decisions that are not only influenced by industry standards but also customized to their specific circumstances, limitations, and goals. This thoughtful approach allows for the selection of cybersecurity solutions that are not only highly regarded in the industry but also truly effective and efficient within an organization's operational framework.

As we move on to the next section, it's important to keep in mind that our ultimate objective is to achieve a strategic alignment between cybersecurity solutions and the needs of the organization, thereby enhancing overall security readiness and operational resilience. This ensures that our journey through the landscape of cybersecurity is both knowledgeable and purposeful, resulting in protection mechanisms against emerging threats.

Thoroughly assessing vendors

When choosing a vendor, it's essential to consider not only their products or services but also the company's history and reputation. To make an informed decision, research the quality of their offerings, specifically ensuring that the vendor's product or service precisely fits your overall strategy and infrastructure. This section will provide the necessary tools and knowledge to evaluate these aspects thoroughly, ensuring that your chosen vendors meet your organization's specific needs and expectations.

When assessing a vendor, examining various aspects of their operations is essential. One crucial area is their track record, which includes their cybersecurity experience, ability to handle threats similar to those faced by your organization, and success stories. A vendor's track record can provide valuable insights into their expertise and reliability, making it a crucial consideration when assessing their suitability.

A vendor's product performance and service quality can be better understood by looking at customer testimonials that provide real-world insights. This feedback often highlights strengths and weaknesses that may not be apparent from product specifications or vendor presentations. Be cautious not to rely solely on testimonials that the vendor shares, but also try to engage directly with these reference customers.

It is also vital to consider post-sale support. Good customer service, issue responsiveness, and resource availability are key factors for effective cybersecurity. A vendor's commitment to supporting their clients post-purchase indicates their dedication to customer satisfaction and product excellence.

Post-sale support is often overlooked when making a technology decision but can cause frustrations and issues if it is lacking.

It is essential to consider how well a vendor's offerings align with your organization's specific requirements. This includes technical capabilities, scalability, integration ease, and compliance with regulations and standards. Considering all these factors, this section will teach you how to conduct a detailed assessment of potential vendors. By developing the skills to conduct thorough vendor assessments, you can make informed decisions that address your current cybersecurity challenges and position your organization well for future security needs.

Evaluating vendor credibility and track record

Begin by researching the vendor's history in the cybersecurity industry. Consider factors such as how long they have been in business, the evolution of their product offerings, and their growth over the years. Vendors that have been in the industry for a long time will likely have a depth of experience and a proven track record. However, consider newer vendors, as they might bring innovative approaches and cutting-edge technology.

Assessing a vendor's track record, case studies, and success stories can help as they provide real-world examples of how the vendor's solutions have been implemented and the outcomes that have been achieved. When reviewing these, consider scenarios that closely align with your organization's challenges.

Industry recognition can also indicate a vendor's credibility and excellence. Look for awards from reputable industry bodies or recognition in well-known cybersecurity publications. Although such recognition should not be the sole basis of your evaluation, it can complement your understanding of the vendor's standing in the industry. Be aware, though, that vendors often purchasing awards so make sure you are aware of the qualification criteria of these awards ahead of time.

To assess the vendor's expertise in handling specific cybersecurity challenges relevant to your organization, examine their technical capabilities, the robustness of their security features, and their experience in industries similar to yours. For instance, if you are particularly concerned about ransomware attacks, evaluate how well the vendor's solutions have managed such threats in the past.

To evaluate a vendor's credibility and track record, there are several practical steps you can take. Utilize online resources, such as the vendor's website, industry forums, and cybersecurity publications, to conduct online research. Access case studies from the vendor's website or request them directly and analyze them with a focus on relevance to your needs. Look for any awards or recognition the vendor has received, which can often be found on their website or in industry reports. Seek informal insights or experiences related to the vendor by engaging with peers in the industry or online forums.

Evaluating a vendor's credibility and track record is crucial in making an informed decision. Thoroughly researching the vendor's history, analyzing their success stories, considering their industry recognition, and understanding their expertise in handling specific cybersecurity challenges can help you choose a reliable and technically capable cybersecurity partner that is well-respected in the industry.

Analyzing customer feedback and post-sale support

Gathering feedback from various sources is essential to fully understanding how a cybersecurity solution performs in different operational contexts. Testimonials and reviews are beneficial as they offer insights into the solution's practicality, ease of use, and effectiveness. When analyzing feedback, it is essential to look for detailed accounts of how the solution met specific security challenges and note any recurring praises or criticisms.

Customer feedback is invaluable for gauging a vendor's solution's real-world effectiveness and user satisfaction. Unlike controlled environments of product demonstrations, customer testimonials and reviews provide insights into how the solutions perform under diverse, real-world conditions. In analyzing customer feedback, it is essential to consider comments on the solution's usability, reliability, and effectiveness, and how it meets specific security needs.

There are various channels to collect customer feedback. Besides the testimonials on the vendor's website, reviews on independent platforms, forums, and industry publications are all valuable. When interpreting feedback, consider the following:

- **Diversity and authenticity**: Look for feedback from diverse customers that suggests a solution's broad applicability. Be wary of reviews that seem overly promotional or lack specific details, as they may not be entirely authentic.

- **Common themes**: Identify common themes in the feedback, such as specific strengths or areas for improvement. This can provide a clearer picture of the solution's performance.

- **Contextual relevance**: Pay attention to feedback from organizations similar to yours in size, industry, or security needs. This can provide more relevant insights for your particular context.

Online platforms and communities are invaluable for gathering candid customer feedback. In addition to the vendor's website and independent review sites, platforms such as Reddit can be particularly insightful. Security professionals often turn to Reddit to seek peer opinions and share experiences of cybersecurity solutions. Subreddits dedicated to cybersecurity are rich with discussions in which real users share their experiences, challenges, and success stories with various solutions. These discussions can provide unfiltered insights into solutions' actual performance and reliability.

To gather feedback, try the following practical steps:

- **Explore independent review sites**: Visit sites such as Gartner Peer Insights or TrustRadius to access unbiased customer reviews.

- **Engage in online forums**: Participate in forums and subreddits where cybersecurity professionals discuss their experiences. Subreddits such as *r/cybersecurity* and *r/netsec* can be particularly helpful.

- **Direct outreach**: Contact other organizations that have used your chosen solutions. They can provide firsthand insights that might be outside of public reviews.

A vendor's post-sale support's effectiveness can significantly impact a cybersecurity solution's long-term success. This includes technical support and regular updates, maintenance, and training resources. Evaluate the support structure's responsiveness, the comprehensiveness of maintenance services, and the availability of educational materials.

Testing the support system during decision-making can provide a clear picture of what to expect. This could involve the following:

- **Contacting support**: Engage with the vendor's support team with specific inquiries or hypothetical scenarios to gauge their responsiveness and expertise.

- **Reviewing update policies**: Examine the vendor's history and policies regarding software updates, focusing on frequency and responsiveness to emerging threats.

- **Assessing training resources**: Look into the training and resources the vendor provides. Practical training can significantly impact how well your team can utilize the solution.

Cybersecurity professionals can make well-rounded decisions by utilizing various resources, from official testimonials to community-driven platforms such as Reddit, and by directly engaging with the vendor's support structure. This approach ensures that the chosen solutions are technically capable and backed by positive real-world experiences and robust support, aligning perfectly with the organization's long-term cybersecurity strategy.

Aligning vendor offerings with organizational requirements

To align vendor offerings with organizational needs, it's crucial to evaluate the technical capabilities of the products. This involves assessing whether the solution's features adequately address your primary security concerns. It's also important to consider whether the solution is adaptable to future cybersecurity challenges.

Ensuring scalability is crucial for organizations that are growing or have fluctuating needs. Choosing a cybersecurity solution that can meet your current requirements and scale up as your organization grows is important. This requires considering both the product's scalability and the vendor's ability to support scaling through flexible licensing options or modular features.

The ease of integration with existing systems is another crucial factor to consider. A solution that requires extensive modifications to your current infrastructure or is incompatible with other vital systems can lead to increased costs and implementation challenges. It is important to assess how the solution can integrate with your existing IT environment.

Ensuring that the cybersecurity solution adheres to relevant laws and regulations is crucial to avoid legal penalties and reputational damage. This involves reviewing the solution's compliance features and the vendor's history of adhering to regulatory standards.

Practical steps for alignment

To ensure that your organization's cybersecurity needs are met, it's important to take practical steps toward alignment. The following steps can guide you toward selecting a vendor whose solutions match your requirements:

1. **Conduct a needs analysis**: Begin by undertaking a comprehensive analysis of your organization's cybersecurity needs. This should include an evaluation of technical requirements, scalability needs, integration capabilities, and compliance obligations.

2. **Match features to needs**: Compare the features of the vendor's solutions directly with the needs identified in your analysis. Prioritize solutions that most closely match your requirements.

3. **Consult with IT teams**: Engage with your IT team to assess the feasibility of integrating new solutions into the existing infrastructure.

4. **Review compliance documentation**: Request detailed compliance documentation from the vendor. Ensure their solutions meet all the regulatory standards and practices relevant to your industry.

5. **Pilot testing**: If feasible, test the shortlisted solutions within your environment. This hands-on approach can provide valuable insights into the solution's integration, scalability, and overall fit.

By evaluating technical fit, scalability, integration capabilities, and compliance alignment, cybersecurity professionals can ensure that the solutions they choose provide robust security and seamlessly integrate into and grow with their organizations. This tailored approach to vendor selection is crucial for establishing a strong, effective, and compliant cybersecurity posture.

Summary

As we conclude our exploration of utilizing analyst and third-party testing reports, we have equipped ourselves with the knowledge and strategies to navigate through the noise of marketing and find cybersecurity solutions that truly match our needs. Throughout this journey, we have learned how to evaluate vendor claims, make sense of external reports, and conduct thorough assessments to ensure that the solutions we choose not only make promises but also deliver practical effectiveness while integrating seamlessly into our operational contexts.

Some key takeaways include the significance of adopting a specific mindset when approaching vendor claims by recognizing the value of reports in validating these claims and understanding the importance of comprehensive evaluations, beyond just technical specifications. Readers have acquired skills in deciphering cybersecurity terminologies and reports evaluating how well solutions can be applied in real-world scenarios, and in involving stakeholders in decision-making processes to ensure a strategic fit.

Leveraging Existing Tools for Enhanced Security

Maintaining cybersecurity measures is crucial for safeguarding organizations in today's evolving digital landscape. In this chapter, we aim to assist cybersecurity professionals in optimizing their current tools and technologies. Our focus is on maximizing the potential of what you already possess.

Throughout this chapter, practical tips and techniques are shared that can be readily applied in real-world scenarios. By the time you reach the end of this chapter, you will have gained an understanding of cybersecurity tools and how to utilize them effectively in safeguarding your organization's digital assets.

In this chapter, we will cover the following topics:

- Identifying existing and required tools and technologies

- Repurposing and integrating tools for enhanced security

- Optimizing tool usage for maximum value

Identifying existing and required tools and technologies

In the world of cybersecurity, it is crucial to have a grasp of the tools and technologies available to effectively manage your defense. This section takes you on a journey of exploring the cybersecurity landscape within your organization. It goes beyond listing tools; it aims to help you confidently understand their roles, capabilities, and how they work together to create a defense strategy. The process starts with conducting an inventory of your cybersecurity arsenal. From antivirus software to threat detection systems, each tool plays a significant role in protecting your digital assets. This inventory isn't a checklist; it provides an evolving overview that reflects changes in both the tools themselves and the cyber threats they combat. We dive into methods for categorizing these tools, ensuring that you have an understanding of their functions and applications. Categorization not only simplifies tool management but also enhances your awareness of areas where your security setup is strong or vulnerable. For example, recognizing that weaker endpoint protections can offset network security measures can guide improvements with confidence. Additionally, we emphasize the importance of assessing how effective these tools are in practice.

In an evolving era of cyber threats, it's important to recognize that tools that were effective yesterday may not be sufficient tomorrow. It is crucial to evaluate these tools against emerging threats to maintain a resilient cybersecurity stance. Additionally, it's essential to understand the significance of identifying gaps in your cybersecurity framework. Having an understanding of what tools are missing or underperforming is critical for strengthening your defenses against threats with a proactive approach. This approach prepares you to tackle security challenges and empowers you with the confidence to anticipate and counter risks.

Cataloging your cybersecurity arsenal

In the realm of cybersecurity, establishing a comprehensive catalog of your cybersecurity tools is a crucial step toward building robust defense measures. This catalog goes beyond being an inventory; it serves as an asset providing clear visibility into your organization's cybersecurity capabilities and readiness. The process of cataloging involves identifying, organizing, and documenting each tool in detail, forming the foundation for a cybersecurity strategy.

The first stage in constructing this catalog involves identifying all cybersecurity tools and technologies in use within your organization. This step requires conducting an examination of your IT infrastructure, including everything from basic antivirus software to complex network security systems and data encryption tools.

For example, a large company might use **global threat intelligence (GTI)** tools and localized firewalls to protect against threats, in regions. On the other hand, a small business may rely on cloud-based security solutions for both cost-effectiveness and scalability. Regardless of their scale or complexity, each tool plays a role in the security framework and requires careful documentation.

The next important step is organizing these tools. This involves categorizing them based on their domain, such as network security, endpoint protection, or data management. Alternatively, they can be categorized based on their importance. For instance, an e-commerce company could categorize its tools into *customer data protection* (such as SSL certificates and encryption tools) and *transaction security* (such as fraud detection systems). This categorization not only helps identify any overlaps or gaps in cybersecurity defenses but also streamlines the management process.

Thorough documentation of each tool is vital. This should include details about its version, configuration settings, customization options, and the IT personnel for its maintenance. For instance, if a healthcare provider uses a **Health Insurance Portability and Accountability Act (HIPAA)**-compliant data encryption tool, they must document information about its configuration settings. The importance of having documentation becomes evident in situations such as data breaches or system failures. It enables efficient responses to such crises.

Large-scale enterprises such as banks or government agencies implement this process in real-world scenarios by maintaining a database of cybersecurity tools. This database includes information such as purchase dates, life-cycle statuses, and compatibility details with systems. It empowers these organizations to swiftly adapt to threats or seamlessly integrate tools.

In summary, regularly cataloging your cybersecurity arsenal is a task that necessitates updates and reviews. As the cyber threat landscape evolves and new technologies emerge, it's crucial to keep your catalog comprehensive and up to date. By doing so, your organization will always be prepared to handle the challenges of the digital world.

Assessing tool effectiveness and relevance

Evaluating the effectiveness and relevance of cybersecurity tools goes beyond merely possessing a variety of them. It involves a nuanced process that requires an in-depth understanding of both the tools themselves and the changing cyber-threat landscape. This subsection explores methodologies for assessing cybersecurity tools to ensure they are operational and align optimally with future security requirements.

The first step in evaluating the effectiveness of a tool is to establish performance metrics. These metrics may include factors such as the rate of detecting false positives, response time, and impact on users. For example, if a company is using an **intrusion detection system** (**IDS**), it would monitor how often the system accurately detects threats versus alarms. It's crucial to avoid a number of positives as it can desensitize security teams and potentially lead to overlooking real threats. The *WannaCry* ransomware attack in 2017 serves as an example where organizations that had up-to-date and properly configured antivirus software were able to prevent the attack. This highlights the significance of tool effectiveness.

Relevance is another aspect when assessing tools. It involves analyzing how well these tools address cyber threats and compliance requirements. For instance, due to the increasing popularity of cloud computing tools, offering security monitoring and management has become more pertinent. An illustrative real-world example can be seen in how many organizations shifted toward cloud-based security solutions following the adoption of work practices during the COVID-19 pandemic. These tools provided the flexibility and scalability to manage security in a distributed work environment.

Furthermore, integration capability is another factor when considering tools for evaluation purposes.

In today's interconnected world, it is crucial for tools to seamlessly integrate and communicate with each other. For instance, a **security information and event management** (**SIEM**) system should effectively work together with existing antivirus software and firewalls to provide threat analysis. Companies such as Cisco and IBM have showcased the power of security systems whereby different tools collaborate to offer a defense against complex cyber threats.

Moreover, it is essential to review and update assessment criteria in order to stay aligned with the evolving cyber landscape. This may involve participating in cybersecurity forums following industry practices and benchmarking against organizations. For example, since the implementation of the **General Data Protection Regulation (GDPR)**, numerous organizations had to reevaluate their data protection measures to ensure compliance, resulting in changes in the data security landscape.

To conclude, evaluating the effectiveness and relevance of cybersecurity tools is a process that requires expertise and a deep understanding of the ever-changing nature of cyber threats and regulatory frameworks. By assessing and adapting our approaches, organizations can ensure that their cybersecurity measures are effective against the rapidly evolving threat landscape.

Identifying gaps and future needs

The evolving cybersecurity landscape presents challenges for organizations striving to stay ahead of threats. It is essential to recognize and address any gaps in the security setup while also considering needs in order to maintain a strong defense. This section will explore strategies for identifying these gaps and implementing cybersecurity measures that are future-proof. Real-world examples will be provided for insights.

The first step in this process involves conducting an analysis of the existing cybersecurity infrastructure to identify any gaps. This analysis includes comparing security measures against industry practices and standards. For instance, a retail company may discover through this analysis that while their **point-of-sale (POS)** systems are secure, their customer data storage lacks encryption, which was famously exploited during the Target data breach in 2013. Conducting such an analysis not only highlights areas that require improvement but also helps prioritize investments in cybersecurity.

Incorporating **threat intelligence (TI)** into the gap identification process is also crucial. By analyzing trends in cyber threats, organizations can anticipate the types of attacks they are likely to face. For example, a financial institution that notices an increase in phishing attacks targeting the banking sector might decide to invest in email filtering and provide employee training to mitigate this risk effectively. TI played a role during the response to the SolarWinds hack in 2020 when affected organizations promptly assessed and strengthened their compromised systems.

Understanding how technological advancements impact cybersecurity is another aspect to consider. The rise of technologies such as the **Internet of Things (IoT)** and 5G networks brings about possibilities for cyberattacks, which means we need to update our defenses. One example is the *Mirai* botnet attack in 2016, where IoT devices were exploited. Many organizations reevaluated their security protocols for IoT after this attack and identified vulnerabilities in their networks that similar threats could exploit.

Regular compliance checks also play a role in identifying gaps. Compliance standards such as GDPR in the EU or HIPAA in the United States often set the minimum requirements for cybersecurity measures. Conducting compliance audits can reveal vulnerabilities. For instance, during a compliance audit, a healthcare provider may discover shortcomings in protecting data, prompting immediate action to address these gaps.

To conclude, identifying gaps and anticipating cybersecurity needs requires effort and awareness. It involves self-reflection, industry knowledge, and staying informed about trends. By conducting gap analyses, monitoring emerging threats closely, and adhering to compliance standards, organizations can adapt their cybersecurity strategies to effectively combat future threats. This foresight is crucial for developing a cybersecurity infrastructure that can tackle today's challenges while being prepared for tomorrow's uncertainties.

Repurposing and integrating tools for enhanced security

In the changing world of cybersecurity, being able to adapt and come up with solutions is crucial. This section focuses on utilizing the potential of existing cybersecurity tools by repurposing them and strategically integrating them. It's about thinking outside the box and transforming what you already have into a more cohesive security arsenal.

The act of repurposing existing tools goes beyond being cost-effective; it demonstrates an organization's flexibility and resourcefulness in the face of evolving cyber threats. By exploring the uses of these tools, we can uncover capabilities and discover new applications. This section will guide you through the process of reimagining your cybersecurity tools, providing examples and strategies that demonstrate how repurposing can significantly enhance your security.

On the other hand, integration focuses on creating a defense system from different elements. In today's interconnected age, the effectiveness of your security posture often depends on how your tools work together. We will delve into methodologies for integrating cybersecurity tools, ensuring they complement each other and provide reinforcement. This approach streamlines your security processes and can close gaps that isolated tools may leave behind.

Lastly, we will discuss how to maximize efficiency by leveraging tool synergy. Having a collection of tools is important. It's even more valuable when they can work together seamlessly to achieve more than what each tool can do individually. In this section, we'll share real-life success stories and practical advice on how to align cybersecurity tools for maximum synergy. Our goal is to provide you with strategies that empower your organization to strengthen its defense mechanisms against the evolving landscape of cyber threats.

Repurposing of cybersecurity tools

In the field of cybersecurity, being able to adapt and think creatively using the resources at hand can make a difference. This section focuses on the use of existing cybersecurity tools, which involves finding expanded ways to improve security measures. It's about thinking beyond their functions and uncovering their potential.

One of the steps in repurposing these tools is understanding their complete capabilities. For instance, consider a network monitoring tool that is typically used to track data flow and identify anomalies. By making adjustments to its settings, it can also be utilized for detecting insider threats. This means it can monitor for data access or transfers that may suggest activity by employees within an organization.

In a real-world scenario, a financial institution adapted its network monitoring tools to identify employee activities, resulting in a reduction in insider fraud.

Another example involves firewalls, which are traditionally employed to protect networks against threats through network segmentation. However, by reconfiguring firewall rules, organizations can restrict the movement of threats within their networks as well. This approach proved valuable for a healthcare provider when they faced an attack as the firewall acted as a barrier, preventing the attack from spreading into critical patient data systems.

Similarly, web filtering tools commonly used to block access to websites can also be repurposed for **data loss prevention (DLP)**. By customizing filters, these tools have the ability to prevent the transmission of information outside of the company's network. To comply with data protection regulations, an e-commerce company successfully implemented this strategy by utilizing its web filters to avoid any leakage of customer data.

Organizations can enhance their threat detection and response capabilities by integrating their system with other security tools such as a SIEM system. For example, a manufacturing company integrated its endpoint security alerts with its SIEM system, resulting in more responses to malware detections across its entire network.

In summary, repurposing cybersecurity tools in different ways requires both expertise and creative thinking. By exploring beyond the applications of these tools and understanding their capabilities, organizations can improve their security posture. This approach can maximize the value of existing investments while also adding an extra layer of resilience against ever-evolving cyber threats.

Integration of security tools

Integration of security tools plays a role in building a strong defense system. It's not about having tools; it's about making sure these tools work together effectively to provide comprehensive and layered security. This section explores the methods and advantages of integrating cybersecurity tools, supported by real-world examples.

The essence of tool integration lies in establishing a flow of information and response mechanisms between security systems. For example, when an IDS is integrated with an SIEM system, it greatly enhances the ability to detect threats. A prominent illustration can be found in the finance sector, where banks have integrated their IDS with SIEM systems to analyze intrusion data in time and respond swiftly and cohesively to breaches.

Another crucial aspect of integration is combining security with network security tools. In one scenario, a healthcare organization unified its endpoint protection platform with its network security solutions. This integration enabled them to isolate devices identified by the endpoint protection, preventing malware from spreading across their network. Such coordination is crucial in environments that handle data and require response times.

Integrating **identity and access management (IAM)** tools with security systems is also crucial. For instance, a technology company successfully merged its IAM solutions with its data encryption systems, guaranteeing that access controls were closely linked to data security protocols. This integration not only enabled access management but also bolstered data protection, which was especially vital considering the company's handling of sensitive **intellectual property (IP)**.

Cloud security is another area where integration can bring benefits. Organizations must be able to seamlessly integrate their cloud security tools with their on-premises security infrastructure. This integration provides a perspective of security across all operations, enabling efficient monitoring and management of potential threats in a hybrid environment.

In summary, strategically integrating security tools is indispensable for a cybersecurity strategy. It ensures that each tool not only fulfills its function but also contributes to a broader, more cohesive security system. By integrating these tools, organizations can achieve a synergy that enhances their ability to detect, respond to, and mitigate cyber threats, thus fortifying their cybersecurity posture and adapting to the changing digital threat landscape.

Maximizing efficiency through tool synergy

The synergy between cybersecurity tools plays a role in establishing an efficient and effective security infrastructure. This section will explore how aligning cybersecurity tools to work together can greatly enhance their combined strength and effectiveness. This strategic alignment not only boosts the capabilities of each tool but also creates a more robust and comprehensive defense mechanism against cyber threats.

Integrating **TI platforms (TIPs)** with security systems is an important aspect of achieving tool synergy. For instance, a global financial services company successfully integrated its TIP with its existing antivirus and firewall systems. This integration enabled real-time threat data to be informed, enabling them to automatically update firewall rules and antivirus signatures, significantly improving the company's ability to proactively block threats. Taking measures is crucial in industries where the consequences of a security breach can be substantial.

Another example of tool synergy involves pairing data encryption with **role-based access control (RBAC)** mechanisms. A financial institution achieved this by synchronizing its encryption solutions with its RBAC system. This setup guaranteed that only staff members with specific roles and permissions could access sensitive financial records, ensuring data remained encrypted and secure despite any unauthorized access attempts.

Synergizing cybersecurity systems can involve utilizing analytics tools across platforms. For example, a retail chain employed analytics tools to correlate data from its network monitoring and POS systems. This synergy enabled them to identify patterns that indicated tampering or data skimming in the POS system, which is a common threat in the retail industry. By connecting the dots across systems using analytics, the company significantly improved its fraud detection capabilities.

Automation also plays a role in achieving tool synergy. A technology company automated the interaction between its **incident response platform** (**IRP**) and network security tools. Whenever a threat is detected, the system automatically initiates containment protocols such as segmenting the network or isolating affected devices. This rapid response minimizes attack spread, a factor in today's fast-paced digital landscape.

To summarize, maximizing efficiency through tool synergy means creating an interconnected and cohesive cybersecurity ecosystem where each component fulfills its function and contributes to security readiness. Organizations can enhance their defense mechanisms by aligning and integrating different tools beyond what each individual part offers. Embracing this approach is essential for combating complex and sophisticated modern cyber threats.

Optimizing tool usage for maximum value

In cybersecurity, the true power of tools lies not in owning them but in utilizing them optimally, unlocking the potential of these tools through usage and ongoing management. This section serves as a guide for professionals looking to make their cybersecurity investments more efficient and effective.

Understanding how to optimize tool usage is crucial. It involves delving into the intricacies of each tool, tuning configurations, and customizing functionalities to perfectly align with an organization's requirements. This process is similar to building a functioning machine where every part operates at its best, contributing to a more resilient whole.

In addition to setup and customization, continuous monitoring and regular audits are essential for maintaining tool effectiveness. This entails implementing mechanisms for tracking performance and periodically evaluating tools against evolving cyber threats and changes within the organization's IT infrastructure. For instance, what might have been considered a configuration a year ago could now be outdated due to types of cyberattacks or shifts in technology.

One critical aspect of optimizing tools is the element of training and knowledge sharing. The effectiveness of advanced tools heavily relies on the expertise of the teams using them.

Therefore, it is crucial to empower employees by equipping them with the required skills and promoting a culture of learning and information exchange. The primary objective is to optimize the utilization of tools and establish a security environment that's responsive, adaptable, and well informed. Upon completing this section, readers will possess strategies for managing their cybersecurity tools, thereby maximizing their value and making substantial contributions to the organization's overall security posture.

Advanced configuration and customization of tools

In cybersecurity, the effectiveness of tools greatly improves when they are finely tuned and customized. This section explores how adjusting cybersecurity tools according to an organization's context can lead to enhancements in security effectiveness. It involves understanding the tools themselves and the unique threats and challenges the organization faces.

The first step in this process is gaining an understanding of the features of each cybersecurity tool in your arsenal. For instance, a network security tool may offer options for threat detection, level of detail in logging, or integration with security systems. A real-life example is observed in institutions where IDS are configured to monitor types of financial fraud. By customizing these settings, these institutions can substantially reduce false positives.

Another important aspect is tailoring security policies within the tools. For example, an e-commerce company might customize its **web application firewall** (**WAF**) by implementing rules for transaction pages compared to parts of its website. This targeted approach enhances security where it is most needed and optimizes site performance by avoiding unnecessary security checks on less sensitive areas.

Customization can also go a step further by integrating security tools with systems within the organization to enhance effectiveness. A prime example is when a corporation's endpoint security solutions are integrated with employee management systems. This integration enables the implementation of dynamic security policies that adjust based on the users' role and location, providing a context-aware approach to safeguarding.

Additionally, advanced configuration often involves automating responses triggered by events. For instance, in a healthcare setting, security tools can be configured with automated scripts that immediately isolate devices showing signs of an infection, thereby halting the spread of the infection to patient data systems. Taking measures becomes crucial in environments where a security breach could have severe consequences.

To summarize, configuring and customizing cybersecurity tools is essential for achieving protection and efficiency. Remember – it is critical to ensure your technology is tailored to your environment. This process requires comprehending an organization's requirements and challenges and understanding its cybersecurity tools' capabilities. By tailoring and fine-tuning these tools, organizations can establish a personalized and highly effective security infrastructure that aligns perfectly with their specific operational environment.

Performance monitoring and regular audits

In the changing world of cybersecurity, monitoring performance and conducting regular audits are crucial to ensure that security tools remain effective and adaptable to new threats. This section explores the significance of these practices and provides real-life examples of how they can enhance cybersecurity defenses.

Continuous performance monitoring forms the foundation of tool management. It involves assessing the capabilities of each tool in detecting, preventing, and responding to security threats. For instance, a multinational corporation might utilize a dashboard to keep track of antivirus software, IDS, and firewall performance. Real-time data on threat detection rates and response times can help identify issues such as outdated signatures or misconfigured settings. An excellent example showcasing this practice is observed in the banking sector, where continuous monitoring of transaction monitoring systems plays a role in detecting and preventing activities.

Furthermore, conducting security audits is another aspect. These audits evaluate the effectiveness of security tools based on cybersecurity standards while pinpointing areas that require improvement. For instance, a tech company may audit its cybersecurity infrastructure encompassing network security, data encryption, and employee access controls. Through these audits, overlooked vulnerabilities such as encryption protocols or excessive access permissions can be identified—vulnerabilities that attackers could potentially exploit.

Incorporating evaluations such as penetration testing can also bring about benefits. These evaluations involve simulating cyberattacks to assess the strength of the security infrastructure. For instance, a healthcare provider may engage a third-party cybersecurity firm to conduct penetration testing on its data systems. The insights gained from these tests can be precious, helping identify vulnerabilities that internal audits might overlook and validating the effectiveness of security measures.

It's also important to extend performance monitoring and audits to include compliance with regulatory standards. Regular compliance audits play a role in sectors such as finance or healthcare, where data protection regulations are stringent. A financial institution might conduct an audit of its security tools to ensure they comply with regulations such as GDPR or HIPAA, guaranteeing that the organization safeguards its data and meets requirements.

In summary, continuous performance monitoring and regular audits are vital for maintaining cybersecurity measures. They provide insights for tuning these measures according to current threat landscapes and compliance requirements. By evaluating and adjusting their security tools, organizations can ensure their defenses remain robust and capable of countering evolving cyber threats.

Training and knowledge sharing

The effectiveness and efficiency of cybersecurity tools are greatly improved through training and knowledge sharing. In this section, we emphasize the importance of empowering the following aspect of cybersecurity – the users and administrators who utilize these tools. We explore how training and knowledge sharing can maximize the value of cybersecurity investments with real-life examples.

An important factor is providing training to staff regarding cybersecurity tools and practices. For instance, a large telecommunications company conducted employee training sessions using its implemented SIEM system. These sessions covered aspects of the SIEM system and its role within the broader context of the company's cybersecurity strategy. As a result, employees became more engaged and proficient in utilizing this system, improving threat detection and response times.

Sharing knowledge also plays a role in optimizing tool utilization. Creating an environment where employees can share insights and experiences about using cybersecurity tools leads to efficient usage. For example, an IT services firm established a platform for knowledge sharing where staff members could exchange tips, best practices, and lessons learned from their experiences with cybersecurity tools. This platform became a resource that fostered employee collaboration while enhancing the firm's security posture.

Tailoring training programs to organizational roles can greatly improve cybersecurity tools' effectiveness. For instance, in a healthcare setting, medical staff receive training on handling patient data and utilizing encryption tools. In contrast, IT staff undergo technical training on network security and data breach response protocols. This role-specific training ensures that every organization member can contribute effectively to cybersecurity efforts.

In addition, mentorship programs and continuous learning opportunities are crucial. A financial institution implemented a mentorship program where experienced cybersecurity professionals offered guidance and support to employees. This program not only accelerated the skills development of staff but also cultivated a culture of ongoing learning and professional growth.

To sum up, comprehensive training and knowledge sharing play roles in optimizing the utilization of cybersecurity tools. They ensure that the human aspect of cybersecurity is well prepared to handle tools. By investing in training initiatives and fostering a culture of knowledge exchange, organizations can significantly enhance the efficacy of their cybersecurity measures, transforming their resources into valuable assets in defending against cyber threats.

Summary

In this chapter, we have delved into an approach to bolstering an organization's cybersecurity position by utilizing the tools at its disposal. Our exploration began with cataloging and evaluating existing cybersecurity tools, emphasizing the significance of understanding what is currently in place and how it can be optimized. This groundwork set the stage for evaluating these tools' effectiveness, ensuring they align with threats, and identifying any gaps in the cybersecurity strategy.

Our attention then shifted toward ways to repurpose and strategically integrate these tools. By thinking outside the box about their functionalities, we discussed how they can be adapted to tackle challenges, enhancing security without incurring expenses. We highlighted the role of integrating tools to create a more coherent and effective defense system where their combined value surpasses their contributions.

Moreover, we explored the importance of optimizing each tool through configuration and customized practices tailored to meet an organization's requirements. This was further supported by emphasizing performance monitoring and regular audits to ensure these tools remain effective and compliant with the evolving cybersecurity landscape.

Lastly, we focused on training initiatives and knowledge sharing within organizations. We explored how providing employees with the required skills and fostering a learning culture can greatly improve the effectiveness of cybersecurity measures. This approach ensures that people are well prepared to use tools.

To sum up, this chapter emphasized that an organization's cybersecurity strength isn't solely reliant on acquiring tools but also on effectively utilizing and maximizing the potential of existing resources. Professionals can significantly enhance their organization's defenses against the changing landscape of cyber threats by assessing, repurposing, integrating, and optimizing cybersecurity.

In the next chapter, you will learn about how to select and implement cybersecurity solutions that fit your organization's needs.

9

Selecting and Implementing the Right Cybersecurity Solutions

In this changing landscape of cyber threats, the real challenge lies not only in understanding a wide range of security tools but also in aligning them precisely with the unique needs and culture of your organization. The goal of this chapter is to equip you with the skills to navigate this terrain while ensuring that your decisions are both strategic and impactful.

Selecting a cybersecurity solution involves more than examining specifications. It entails evaluating how these solutions integrate into your existing IT infrastructure, how they will adapt as your organization evolves, and their effect on day-to-day operations. You will learn how to tackle challenges such as compatibility issues, budget constraints, managing changes, and meeting compliance. This chapter goes beyond technology; it focuses on integrating people, processes, and tools to nurture a cybersecurity culture.

Furthermore, we'll delve into the implementation and integration of these solutions. Implementing cybersecurity measures is a journey rather than a one-time destination. You will explore developing implementation plans, initiating training programs and awareness campaigns, and establishing feedback mechanisms for continuous improvement. Through real-world examples, you will gain insights, into strategies and their outcomes – equipping you to effectively bolster your organization's cybersecurity posture.

In this chapter, we will cover the following topics:

- **Understanding the threat landscape**: This section focuses on the importance of recognizing the variety of cyber threats that organizations face, emphasizing the need for a tailored cybersecurity strategy that addresses specific vulnerabilities and threats unique to different industries

- **Assessing system compatibility and integration**: This section delves into the necessity of ensuring that cybersecurity solutions are compatible with existing IT infrastructure, discussing the evaluation of new tools for seamless integration, scalability, and compliance with industry standards

- **Scalability and future-proofing cybersecurity solutions**: This section covers the importance of choosing cybersecurity solutions that can adapt and grow with the organization, highlighting the need for scalable, adaptable, and cost-effective security measures that support future organizational growth

- **Compliance and industry standards in cybersecurity solutions**: This section emphasizes the critical role of compliance with laws, regulations, and industry standards in cybersecurity, outlining steps to evaluate solutions for compliance and integrate these requirements into a comprehensive cybersecurity strategy

- **Best practices for selecting security tools**: This section offers guidance on conducting market research, involving stakeholders in the selection process, and evaluating the cost-effectiveness and ROI of cybersecurity solutions, advocating for a thorough approach to choosing the right tools and vendors

- **Implementing and integrating cybersecurity solutions**: This section discusses the process of implementing cybersecurity solutions, from planning and user training to monitoring, maintenance, and regular updates, stressing the importance of a strategic implementation plan, user adoption, and establishing a feedback loop for continuous improvement

Factors to consider when selecting cybersecurity solutions

Now, let's shift our focus to the factors that should guide your decision-making when choosing cybersecurity solutions. This step is crucial for protecting your organization's assets. It requires a thorough understanding of both your internal environment and the external threat landscape. It's not randomly selecting a tool; it's about finding a solution that suits your specific requirements, aligns with your existing systems, and can adapt to future changes.

To begin with, it is essential to have an understanding of the threats your organization faces. Cyber threats can vary greatly across industries and even among organizations. Therefore conducting a comprehensive threat analysis is vital. This will help you identify not only the types of attacks you are most vulnerable to but also understand their impact. Armed with this knowledge, you can choose a solution that protects against these identified risks.

Furthermore, any cybersecurity solution should seamlessly integrate with your IT infrastructure. This involves considering how new tools will fit into your existing systems and processes and how they will be received by those who use them daily. Scalability and compliance are also factors to consider.

The ideal solution should not only meet your needs but also have the flexibility to grow alongside your organization.

It is important to follow the established norms and regulations of the industry, which will ensure that your cybersecurity measures are both efficient and, in accordance, with the law.

Understanding the threat landscape

It is essential to understand the threats that exist to choose effective solutions. This deep understanding forms the foundation of a cybersecurity strategy ensuring that defenses are not only robust but also tailored to address specific threats faced by an organization.

The constant evolution of cyber threats

One factor to consider is how cyber threats continually evolve. These threats are driven by advancements and changes in attacker tactics and global socio-political developments. For example, we have witnessed a surge in attacks targeting access systems. A significant incident was when a major corporation experienced a significant data breach through their working platform highlighting the need for solutions such as enhanced multi-factor authentication and secure **virtual private networks** (**VPNs**) to protect remote access points.

Threats unique to different industries

The financial sector often falls victim to phishing schemes and ransomware attacks while healthcare organizations are more susceptible to data breaches involving patient information. A notable case was when a prominent hospital encountered an attack that severely disrupted its record systems and demanded a substantial ransom payment. This event highlights the importance of cybersecurity solutions that are specifically designed for industries. For healthcare, it is crucial to have systems for detecting and neutralizing malware threats at an early stage.

Assessing past incidents

Examining security incidents within your organization and among peers in your industry is highly valuable. This retrospective approach helps identify areas to improve and learn about potentially new methods used in attacks. For example, let's consider a company that experienced a compromise of their point of sale systems through a phishing attack. Analyzing incidents can lead to adopting solutions with email filtering capabilities and implementing comprehensive employee training programs to recognize and respond effectively to phishing attempts.

Global and local threat intelligence

Staying updated on both local and global threat intelligence is essential. Cyber threats often exhibit patterns that can be identified through intelligence networks. However, local intelligence is equally important as it provides insights into threats in specific regions. For instance, if a company operates in an area known for state-sponsored cyber espionage, it would greatly benefit from using intrusion detection and encryption technologies tailored specifically to these types of threats.

Developing a customized cybersecurity strategy

The final step in comprehending the threat landscape involves creating a cybersecurity strategy that is tailored to your organization's needs. This strategy should address existing threats while remaining flexible enough to adapt to risks. One example is a technology company that recognized the increasing danger of AI-powered cyberattacks and invested in state-of-the-art cybersecurity solutions based on AI. These tools allowed the company to better anticipate and counter the attacks.

It is also important to consider nonemerging threats, especially with the growing interconnectedness of devices New vulnerabilities are constantly emerging, as seen with a DDoS attack caused by compromised devices, which disrupted internet services across a wide area. This incident highlights the need for cybersecurity solutions that go beyond computer networks and encompass IoT and mobile devices.

Employee awareness and training also play a critical role in understanding and mitigating cyber threats. Human error often serves as an entry point for attacks. For instance, an employee unknowingly leaked data due to falling for a phishing scam emphasizing the importance of cybersecurity awareness programs. Such initiatives can significantly reduce the risk of cyber attacks originating from within an organization.

Collaboration and information sharing between organizations are also crucial. By exchanging experiences and strategies, companies can collectively improve their understanding of the threat landscape. Take, for instance, industries such as banking that have come together to exchange information about threats. This collaboration has significantly enhanced their ability to defend against cyber attacks as a front.

Being aware of the evolving threat landscape requires a new approach. It involves staying up to date with emerging threats, understanding industry risks, learning from past incidents, and having access to both global and local intelligence. Armed with this knowledge, organizations can choose cybersecurity solutions that not only work effectively but fit within their unique context. Real-world examples showcase the impact of this knowledge empowering organizations to make decisions on cybersecurity that are both strategically sound and operationally feasible. It's important to note that this understanding isn't a one-time task; it should be an ongoing effort vital for maintaining a cybersecurity stance in our changing digital world.

Assessing system compatibility and integration

When it comes to choosing cybersecurity solutions, one of the factors to consider is ensuring that they are compatible with your existing systems and can be seamlessly integrated. To achieve this, it is important to evaluate how these new security tools will work within your IT infrastructure. In this section, you will learn the steps and questions that you can ask vendors or include in your **request for proposal** (**RFP**) to ensure that the solutions you select will integrate effectively with your systems.

Understanding your IT environment

To start, it is recommended to conduct an audit of your present IT setup. This should involve documenting all hardware, software, and network configurations, as well as any legacy systems that may still be in use. For example, if a company decides to transition to a cloud-based solution without considering its compatibility with legacy systems, integration issues may arise. To prevent such challenges, it is important to create an inventory of all your IT assets and identify any potential compatibility issues.

Evaluating compatibility with existing systems

When evaluating cybersecurity solutions, it becomes crucial to assess their compatibility with your existing systems to ensure integration and optimal functionality. This assessment involves examining operational aspects. Firstly, it is important to evaluate the software and hardware requirements of the solution against your IT infrastructure. This includes checking whether there are any operating system requirements or demands for processors, memory capacities, or storage needs, from the solution itself. Moreover, understanding how the new solution will integrate with your existing network architecture holds importance.

Will it seamlessly integrate with your network setup? Will it necessitate substantial modifications that could result in downtime or compromised performance? Additionally, consider the compatibility of data – can the new solution effortlessly accommodate your existing data formats and structures? Will it require data conversion or migration efforts? Another crucial factor is its alignment with existing security protocols and compliance requirements. The new solution should not only bolster your cybersecurity stance – it should also adhere to industry standards and regulatory obligations followed by your organization. By ensuring this level of compatibility, you can avoid disruptions and maintain a robust, cohesive, and compliant cybersecurity framework. To achieve this, you should pose questions to vendors. Include them in your RFP. Some key inquiries may include the following:

- What are the minimum hardware and software prerequisites for implementing your solution?
- Can your solution seamlessly integrate with our existing network infrastructure without compromising performance?
- How does your solution handle data from our systems?

Addressing challenges posed by legacy systems

Legacy systems, although often crucial for an organization's operations, can present integration challenges when introducing cybersecurity solutions. The first step is to identify any conflicts that may arise between the existing legacy systems and the proposed security technologies.

To address this, we need to analyze the infrastructure. This involves examining hardware limitations, software dependencies, and network architecture. By understanding these elements, we can identify any conflicts that may arise in terms of system requirements, data formats, or communication protocols.

Once these conflicts are identified, it is crucial to develop a plan that tackles them head-on. This plan may involve updating components of the existing systems to ensure compatibility or gradually replacing legacy systems with modern solutions. Such an approach not only helps mitigate cybersecurity risks but also improves operational workflow by enhancing overall security and performance.

Here are some important questions to consider:

- How does your proposed solution integrate with legacy systems or software?
- What strategies do you recommend for updating legacy systems to ensure compatibility?
- Can you provide case studies or examples where you have successfully integrated with legacy systems?

For an integration process, it is essential to develop an integration plan that includes pilot testing and phased rollouts. Start by identifying areas or departments within your organization that can serve as testing grounds for the systems. This pilot testing phase is crucial as it allows us to identify any issues and ensures compatibility with the existing infrastructure before proceeding with a full-scale deployment.

Based on the insights and feedback we've gathered from these tests, you can refine the integration strategy to better suit the needs of your organization. After that, it is recommended to adopt a phased rollout approach gradually implementing the solution, across parts of your organization. This approach allows for monitoring the impact on operations minimizing disruption to activities and allows you to make incremental adjustments as you extend the system to wider areas of your organization. Planning and executing this integration strategy will not only ensure a smoother transition but also foster greater acceptance and adaptability among users and stakeholders. To ensure integration, here are some questions to consider:

- What is your recommended approach for deploying your solution in a multi-vendor environment?
- How do you support pilot testing and phased rollouts to minimize operational disruption?
- Can you provide an integration timeline that includes milestones?

These questions, when directed to the vendor, will guide you in integrating the new cybersecurity solutions into your existing infrastructure, ensuring a smooth transition and effective security enhancement.

Scaling for future growth

Choosing cybersecurity solutions that can scale with your organization's growth is crucial for long-term cybersecurity strategy. It's important to select tools and systems that are not only effective at your size but also can accommodate future expansion, whether it's increased data volume, additional users, or an expanded network infrastructure.

Ensuring future-proof security solutions is essential for the growth of your organization. You don't want to be stuck with inadequate security measures that expose you to risks. Look for options that can scale as your organization expands, such as add-ons, customizable features, and integration capabilities with emerging technologies. This could involve considering cloud-based security services for their elasticity or choosing software with licensing models that allow expansion. By planning for scalability, you safeguard your organization's future. Ensure that your cybersecurity infrastructure grows alongside your business supporting its development rather than hindering it. Here are some questions to include in your RFP regarding scalability:

- How does your solution handle the increase in data volume and network size?

- What are the associated costs of scaling up your solution?

- Can you provide examples of how your solution has scaled in organizations?

Maintaining operational efficiency is crucial when implementing new security measures. It's important to strike a balance between security and minimal impact on system performance and user experience. Choose cybersecurity solutions that not only effectively thwart threats but also optimize system resources and user workflows.

For instance, a company could choose a security solution that utilizes algorithms designed to protect without compromising system performance. At the time, it is crucial to prioritize user experience by ensuring that security protocols are easy to use and don't overly hinder employee productivity. Take the example of a retailer who could implement a streamlined customer authentication process, ensuring enhanced security without causing significant inconvenience for users. By considering these aspects, organizations can ensure that their cybersecurity measures enhance system performance and the overall user experience of hindering them. When engaging with vendors to evaluate their cybersecurity solutions, it's crucial to understand their impact on your systems and user experience. Ask the following questions to ensure their offerings align with your performance and usability expectations:

- How will your solution impact our systems' performance?

- How do you optimize your solution to minimize any impact on user experience?

- Can you share performance benchmarks or case studies?

Maintaining a balance between security and user-friendliness is crucial. Striking this balance ensures that cybersecurity measures effectively safeguard against threats while being embraced by users and seamlessly integrated into operations. Organizations need to prioritize security without causing inconvenience or disruptions for employees and customers. This can be achieved through the implementation of security protocols and user-friendly authentication methods.

For example, **multi-factor authentication** (**MFA**) can improve security while still being user-friendly by offering options such as recognition or one-time passcodes sent to devices. It is also crucial to have communication and provide user training to ensure that everyone understands and follows the security measures. This helps minimize any resistance from users. By creating a culture that values both protection and user convenience, organizations can establish a cybersecurity environment that's effective and well-received by all stakeholders. When evaluating the user-friendliness and adaptability of cybersecurity solutions, consider asking vendors these critical questions:

- How easy is your solution for nontechnical staff to use?
- What kind of training and support do you offer to ensure simplicity?
- Can the user interface be customized according to our organization's needs?

These questions will help you assess whether a cybersecurity solution aligns with your organization's operational dynamics and user capabilities, ensuring a smoother integration and adoption process across all levels of your workforce.

Implementing testing and feedback mechanisms

Testing and gathering feedback are essential for integration. In the changing world of cybersecurity, continuous assessment and improvement are crucial in making sure that integrated security solutions effectively address evolving threats and organizational requirements. Organizational periodic testing allows you to identify vulnerabilities or inefficiencies in the integrated systems. This may involve penetration testing to identify weaknesses or simulated cyberattack scenarios to assess how well the security measures respond. Important is to collect feedback from users and stakeholders who interact with these integrated systems daily.

To effectively uncover usability issues, identify restrictive security measures, and address new threat vectors, it is crucial to gain insights from routine testing and feedback. This iterative process plays a role in creating a cybersecurity environment where adjustments and enhancements can be promptly implemented ultimately strengthening the overall security posture of the organization.

Handling feedback and requests for system adjustments should be done in a way that promotes communication and collaboration. Moreover, providing support and updates post-integration is essential for maintaining an environment.

To assess system compatibility and integration successfully, it is important to have an understanding of your IT environment. Conducting evaluations of solutions while strategically planning their integration will help strike a balance between security needs and operational efficiency. By asking questions and including requirements in your RFP, you can ensure that the chosen cybersecurity solutions not only safeguard your organization but also seamlessly integrate with your existing infrastructure. This approach is crucial in maintaining an efficient IT environment as cyber threats continue to evolve.

Scalability and future-proofing cybersecurity solutions

To ensure long-term protection, it is essential to have the ability to scale and adapt to challenges. In this section, we will focus on selecting cybersecurity solutions that not only meet your needs but also have the potential to grow and adapt to future threats and organizational growth. Real-world examples will be used to illustrate these concepts along with steps and important questions that should be asked when considering technologies from vendors.

Understanding scalability in cybersecurity

Scalability in cybersecurity refers to the capacity of a solution to expand and adjust along with your organization. To apply this knowledge, it is important to evaluate your cybersecurity measures and identify any limitations in handling increased data volume, traffic, or advanced threats. For instance, a growing e-commerce business initially implemented a cybersecurity setup and later upgraded it to a more robust system as its customer base and transaction volume expanded.

Assessing future needs

It is crucial to assess both your current and future cybersecurity needs. This involves predicting your company's growth trajectory and technological advancements as the threat landscape evolves. Start by conducting an assessment of your existing cybersecurity infrastructure, policies, and practices. Understand the vulnerabilities that currently exist alongside identifying areas for improvement. This includes evaluating the effectiveness of your security solutions and assessing the expertise of your cybersecurity team while ensuring alignment with industry standards and regulatory requirements.

Consider the growth of your organization. How do you expect it to evolve in the coming years? Evaluate expansions in terms of employee numbers, customer base, geographical reach, and digital assets. It's important to stay informed about emerging technologies that may affect your cybersecurity needs. The adoption of technologies such as devices, cloud computing, or AI-driven solutions might require adjustments to your cybersecurity strategy. Also, make sure you keep track of any changes and compliance requirements in your industry as they can significantly influence your cybersecurity demands. When selecting cybersecurity solutions, prioritize those that offer scalability features.

To stay ahead of emerging threats in the cyber world, cybersecurity professionals need to update their knowledge about the threats, vulnerabilities, and attack methods. This involves attending conferences, webinars, and training programs dedicated to cybersecurity. Keep yourself updated on advancements well. Understand how new technologies, such as AI, IoT, and cloud computing, can impact the security landscape within your organization.

Recognize the security risks associated with these technologies and adjust your strategy accordingly. For instance, a manufacturing company acknowledged the increasing threat of attacks and invested in solutions that offer advanced IoT security features.

Consider the cost implications of scalability. Take into account not only the investment but also the long-term cost-effectiveness. In one case, a retail chain opted for a SIEM system, which had upfront costs but resulted in savings over time by reducing the need for frequent upgrades.

When evaluating vendors, ask them about their solution's ability to handle growing data volumes and network expansion. Inquire if their system can integrate with advancements. Additionally, understand the cost structure associated with scaling up their solution.

Integrate scalability into your security policies by updating them to accommodate growth and address threats. A logistics company's periodic policy reviews enable them to adapt to changing scales of operation and emerging threats.

As your workforce expands, ensure that your training programs can scale accordingly. A multinational corporation implemented a training program that could be easily scaled to provide cybersecurity awareness across its growing workforce.

Establish vendor partnerships with companies that understand and support scalability to foster relationships based on growth and development. A technology company partnered with a vendor to gain access to solutions and updates, which helped them grow and adapt to security risks.

It is important to test and evaluate the scalability of your cybersecurity measures. For instance, an online business conducted security audits twice a year to identify areas that needed scaling.

Choosing cybersecurity solutions that are scalable and future-proof is crucial for maintaining security and operational efficiency. By understanding scalability, assessing future needs, selecting solutions, preparing for emerging threats, considering costs, updating policies, providing training at scale, forming partnerships with vendors, and conducting regular evaluations, you can ensure that your cybersecurity infrastructure is robust enough to support your organization's growth while effectively addressing evolving challenges. These steps, along with asking questions when engaging vendors, offer an approach for cybersecurity professionals who seek scalable and future-proof solutions.

Compliance and industry standards in cybersecurity solutions

Making sure you follow the rules and regulations and keeping up with industry standards is not just something you have to do but it can also give you an advantage strategically. This section is here to help you choose cybersecurity solutions that not only meet compliance standards but also improve your security. It includes real-life examples, practical steps, and important questions to ask vendors so that you can effectively apply this knowledge.

Understanding the importance of compliance

In the field of cybersecurity, compliance means sticking to laws, regulations, and guidelines that are meant to safeguard data and privacy. This is crucial to avoid penalties and maintain the trust of your customers. For example, a financial services company was hit with fines because they didn't comply with GDPR due to their cybersecurity solution not adequately protecting customer data. To make use of this knowledge, start by identifying all the rules and standards that apply to your industry and location.

Evaluating solutions for compliance

When assessing cybersecurity solutions, make sure they meet these compliance requirements. That means checking if the solution can handle data securely, reporting any breaches promptly, and providing all documentation for compliance audits. For instance, a healthcare provider working under HIPAA regulations chose a solution that offered data encryption and access controls while also keeping records of data access – which were essential – for their compliance needs.

Aligning with industry norms

In addition to meeting obligations, aligning with industry standards such as ISO 27001 or NIST frameworks can significantly enhance your cybersecurity strategy. For instance, a technology start-up embraced the ISO 27001 standard, enabling them to establish a security management system that enhanced their credibility and fostered customer trust.

Here are some questions for vendors regarding compliance and standards:

- How does your solution guarantee compliance with regulations such as GDPR, HIPAA, or PCI DSS?

- Can your solution furnish reports and logs for compliance audits?

- Does your product adhere to industry standards such as ISO 27001 or NIST?

Incorporating compliance into your cybersecurity strategy

Integrating compliance into your cybersecurity strategy encompasses more than selecting the solution. It necessitates an approach encompassing policy development, employee training, and regular audits. A multinational corporation implemented a strategy that integrated GDPR compliance into every facet of their operations from data handling to employee training.

Frequent updates and audits

Compliance requirements and industry standards are dynamic and evolve. Ensure compliance by updating policies and conducting audits. For example, a retail company established a routine where they reviewed their compliance status bi-annually, adjusting policies and systems in response to regulations and emerging threats.

Collaborating with your cybersecurity solution vendors is crucial to maintaining compliance. By working with them, you can stay up to date with the compliance trends and make necessary adjustments to your solutions. For instance, a finance company established a partnership with their security vendor, who assisted them in navigating the complexities of PCI DSS compliance. Together, they ensured that their payment systems were secure and met all standards.

It is essential to train and raise awareness among your staff regarding compliance requirements. Employee actions have an impact on compliance, especially when it comes to handling data and respecting privacy. An educational institution conducted training sessions for its staff on FERPA compliance emphasizing data handling practices within an academic environment.

Selecting cybersecurity solutions that comply with requirements and align with industry standards is vital for compliance, customer trust, and overall security effectiveness. By understanding the importance of compliance, evaluating solutions based on their capabilities in meeting these requirements, aligning with industry standards, integrating compliance into your strategy, conducting updates and audits, collaborating closely with vendors, and emphasizing training and awareness among employees, you can ensure that your cybersecurity measures are not only legally compliant but also robust and reliable. These steps provide an approach for cybersecurity professionals seeking to navigate the complexities of compliance in today's world.

Best practices for selecting security tools

This section aims to provide a guide for navigating the vast cybersecurity technology market. It ensures that the solutions you choose align perfectly with your organization's needs and challenges.

Selecting the right cybersecurity solution is a process that requires an understanding of the current market landscape. As the range of cybersecurity products and services continues to grow, it becomes crucial to conduct market research. This research will help you grasp the strengths, weaknesses, and suitability of options. In this section, we will walk you through this process by demonstrating how to compare and contrast solutions based on their features, effectiveness, and compatibility with your existing systems.

Involving stakeholders from departments within your organization is another vital step in this process. Cybersecurity affects every aspect of your business; it's not limited to the IT department. Therefore, gathering insights, requirements, and feedback from teams, including finance, human resources, and operations, among others, is crucial. This ensures that the selected solution not only secures your assets but also supports and enhances overall business operations.

At the core of any cybersecurity strategy lies risk assessment and management.

In this section, we'll delve into the process of identifying the risks that your organization may face and explore cybersecurity solutions that can help mitigate these risks. By aligning your cybersecurity measures with your organization's risk profile, you can ensure that your resources are directed towards the most critical areas thereby enhancing the overall effectiveness of your security approach.

Moreover, it's crucial to evaluate the cost-effectiveness and **return on investment (ROI)** of cybersecurity solutions. We understand that every organization has budget constraints and it's important to justify cybersecurity expenditures. This segment of the chapter will guide you in analyzing the costs associated with solutions, estimating their ROI, and striking a balance between considerations and robust security measures.

Conducting comprehensive market research

To ensure protection for your organization, it is crucial to have an understanding of the market landscape and assess potential solutions. This section provides in-depth guidance and real-world examples to illustrate how to conduct market research.

Understanding the landscape of the cybersecurity market

To begin with, it is important to familiarize yourself with the cybersecurity landscape. This involves identifying trends, such as the increasing use of AI in threat detection or the growing prevalence of attacks. For instance, a logistics company noticed a rise in attacks within their industry. They initiated their market research by focusing on solutions designed to combat threats. This led them to explore endpoint protection platforms that are known for their effectiveness against ransomware.

Comparing different cybersecurity solutions

When comparing solutions, it is essential to examine their features in detail. Consider factors such as the type of protection offered (for example, firewalls, antivirus, and others) deployment options (cloud-based or on-premises), and user-friendliness. For example, a law firm required a solution that provided both data encryption and loss prevention. They evaluated products based on these features. They also took into account user interfaces since they had non-technical staff.

When evaluating vendors, it's important to research their background and customer service track record, and how they handle updates and patches. For example, a nonprofit organization, without an in-house IT team chose a vendor known for providing updates and strong customer support. They made this decision based on the vendor's track record in supporting organizations of size.

To stay ahead of the curve, it's crucial to analyze trends and developments. This can involve subscribing to cybersecurity newsletters, attending webinars, and participating in industry forums. Take the case of an institution that focused on the increasing threat of mobile banking attacks. They researched to find solutions that offer mobile security features.

Considering customization and integration capabilities is vital when selecting a cybersecurity solution. A manufacturing company for instance needed a solution that could seamlessly integrate with their existing hardware systems. They prioritized solutions that provided custom integration services so they could enhance their cybersecurity measures without revamping their infrastructure.

User reviews and feedback play a role in understanding how cybersecurity solutions perform in real-world scenarios. It provides insights into their application.

An example of this is an e-commerce business analyzing a range of user reviews to understand how different web security solutions perform during periods of traffic. This is particularly crucial for their website, which experiences traffic.

Finding the balance between cost and effectiveness is a process when it comes to cybersecurity solutions. It requires looking beyond the price and considering the long-term return on investment. A start-up, even though they have budget constraints, decided to invest in an expensive solution that offers comprehensive security features and scalability. They saw it as an investment for their growth.

Getting opinions and seeking consultations from cybersecurity professionals can bring clarity and guidance when selecting the solution. For example, a healthcare provider sought the assistance of a cybersecurity consultant to navigate the intricacies of compliance. They aimed to find a solution that meets their security needs and also that which aligns with healthcare regulations.

Conducting market research in the field of cybersecurity involves delving into the current landscape by carefully comparing different solutions, evaluating vendors extensively, staying up to date with trends, considering integration and customization requirements, paying attention to user feedback, assessing cost-effectiveness, and consulting with industry experts.

By adhering to these guidelines, companies can make informed choices when it comes to cybersecurity solutions. They can select options that not only tackle their security concerns but also position them favorably against future threats and technological progress.

Involving key stakeholders in the selection process

It is essential to involve individuals from different departments to ensure the success of cybersecurity solution selection. This section delves into the practices for engaging stakeholders in this process providing real-world examples and valuable insights that can help apply the acquired skills.

Understanding stakeholder roles

The initial step involves comprehending the employees and responsibilities of stakeholders within your organization's cybersecurity ecosystem. These stakeholders typically consist of IT professionals, management teams, end users, and even customers. For example, when a bank includes its IT team, branch managers, and customer service representatives in the decision-making process for selecting a cybersecurity solution, it ensures an approach that addresses technical, operational, and customer-centric aspects.

Early engagement with stakeholders

Involving stakeholders in the process is vital. This enables an understanding of their concerns and expectations regarding cybersecurity solutions.

In the stages of selecting a data protection solution, an e-commerce company actively involved its marketing and sales teams. This collaborative approach aimed to address their concerns regarding customer data security and privacy.

To ensure communication throughout the process, it was crucial to maintain meetings, updates, and feedback sessions with stakeholders. By keeping everyone informed and involved, the company established a cybersecurity forum in the healthcare industry as an example. This platform allowed stakeholders to openly discuss solutions, share concerns, and provide feedback. Consequently, this inclusive approach led to a decision-making process that considered a range of perspectives.

When choosing cybersecurity solutions, it is crucial to align them with business objectives. This ensures that these solutions support business operations rather than hinder them. For instance, a manufacturing company prioritized production as its goal when selecting a cybersecurity solution. They opted for a security solution that did not adversely affect production line efficiency.

Considering the user experience and training needs of stakeholders is essential for the adoption of cybersecurity measures. A user-friendly solution increases the likelihood of implementation within an organization. For example, an educational institution took into account the proficiency levels among its staff and students when selecting a cybersecurity solution. They chose one with an interface accompanied by training materials.

Balancing opinions during decision-making can be challenging yet vital for the implementation of cybersecurity measures. Striving toward consensus allows for inclusivity in decision-making processes while considering perspectives.

Including stakeholders in the evaluation of cybersecurity solutions is a process that necessitates meticulous planning, effective communication, and a collaborative approach. It is important to engage stakeholders such as customers, partners, and regulatory bodies to ensure fairness and transparency in decision-making. For instance, a financial services firm incorporated customer feedback into its cybersecurity solution selection process to meet its expectations regarding data security and privacy.

Documenting and reviewing stakeholder input is crucial for reference and accountability. This documentation helps track requirements decisions made and the underlying rationale. To illustrate this point, a government agency diligently recorded all stakeholder meetings and feedback during the selection of a security solution. This proved valuable for auditing purposes as compliance.

Furthermore, ongoing engagement with stakeholders after the selection process is vital. Regular involvement during implementation facilitates improvement and adaptability. A retail chain exemplified this by establishing a stakeholder committee that regularly assessed the performance of their implemented cybersecurity solution providing insights for ongoing enhancements.

In summary, involving stakeholders in cybersecurity solution selection requires planning, transparent communication, and collaboration throughout the process. To ensure the adoption of cybersecurity solutions within an organization, it is crucial to consider the needs and viewpoints of different stakeholders. By aligning these solutions with the business objectives, taking into account opinions, involving parties as necessary, documenting decisions, and maintaining ongoing engagement, organizations can choose and implement cybersecurity measures that are not only technically robust but also widely supported and effective across all areas of the organization. This inclusive approach guarantees that the selected solutions are practical, user-friendly, and in harmony with the goals and functioning of the organization.

Performing risk assessment and management

When it comes to choosing cybersecurity solutions, conducting a risk assessment and management is essential. This section delves into the methodologies of risk assessment and its role in the decision-making process for selecting cybersecurity solutions. It provides real-world examples and practical advice to ensure the skills acquired can be effectively applied.

Understanding risk assessment in cybersecurity

In the realm of cybersecurity, risk assessment involves identifying threats and vulnerabilities while evaluating their impact on your organization. It is crucial to start by comprehending which assets require protection and the specific threats they may encounter. For instance, a financial institution might prioritize safeguarding customer data and financial transactions focusing on solutions that offer encryption and fraud detection capabilities.

Prioritizing assets

The initial step in conducting a risk assessment is prioritizing assets. This includes assets such as servers and databases, as well as intangible assets such as company reputation. For example, a retail company recognized its customer database as a high-priority asset and concentrated on solutions that provided data protection measures along with breach detection capabilities.

Analyzing threats and vulnerabilities

Once assets have been identified, it is necessary to analyze threats and vulnerabilities. This entails understanding the types of attacks that could occur and how they may exploit weaknesses within your system. For instance, a healthcare organization investigated security risks, such as ransomware and insider threats. This analysis helped them identify solutions that offer both internal and external security measures.

Evaluating the impact of potential risks

Assessing the impact of risks involves examining the consequences of security breaches. This includes considering losses, damage to reputation, and legal ramifications. To safeguard against data breaches, a technology company invested in cybersecurity solutions that not only prevented but also provided rapid response capabilities.

Strategies for managing risks

Developing risk management strategies entails determining how to handle identified risks. This can involve avoiding, mitigating, transferring, or accepting risks. When faced with the risk of DDS attacks, an e-commerce platform chose to mitigate this threat by implementing DDoS solutions.

Engaging stakeholders in risk assessment

Involving stakeholders in the risk assessment process ensures an understanding of risks from different perspectives. An educational institution engaged faculty members, IT staff, and administration in their risk assessment efforts. This collaboration provided a view of potential security threats and their impacts.

Regularly updating risk assessments

Given that cyber threats are constantly evolving, it is crucial to update risk assessments. A media company has implemented a policy of conducting risk assessments twice a year to proactively address emerging threats and adapt its cybersecurity strategy as needed.

Finding the balance between managing risks and working within budget constraints can be challenging. It's crucial to prioritize risks and allocate resources where they are most necessary. A non-profit organization with funds focused their investment on cybersecurity solutions that targeted their critical vulnerabilities.

Leveraging the insights gained from your risk assessment is key when selecting cybersecurity solutions. This ensures that the chosen solutions effectively address the risks faced by your organization. For example, a logistics firm used the results of their risk assessment to choose a solution that protected against supply chain attacks, which they identified as their risk.

Proper documentation of the risk assessment process is essential for compliance purposes and future reference. It helps demonstrate diligence and informs risk assessments. A manufacturing company diligently documented its risk assessment process, which later proved invaluable during an industry compliance audit.

Conducting risk assessments and effectively managing them are important aspects when it comes to selecting suitable cybersecurity solutions.

To effectively safeguard your organization against its risks, it is crucial to comprehend the vulnerabilities it faces. This involves identifying and prioritizing assets, analyzing threats, evaluating their impacts, involving relevant stakeholders, regularly updating assessments, finding a balance between risks and budget limitations, and using these assessments to inform your choice of cybersecurity solutions. By adopting this approach, you not only strengthen your organization's cybersecurity measures but also facilitate informed decision-making and efficient allocation of resources.

Evaluating cost-effectiveness and ROI in cybersecurity solutions

When it comes to choosing cybersecurity solutions, it is important to consider their cost-effectiveness and ROI. This section delves into the intricacies of considerations in the field of cybersecurity providing a guide on how to assess and make informed decisions based on cost effectiveness and potential ROI. Real-world examples are included to ensure understanding and application of these concepts.

Understanding the financial aspects of cybersecurity

The financial aspect of cybersecurity encompasses more than the purchase price of a solution. It also includes long-term costs such as maintenance, updates, training, and potential expenses in case of a breach. For instance, a small business owner not only considers the costs of acquiring a new firewall but also takes into account the ongoing expenses associated with its upkeep and keeping it up to date.

Analyzing total cost of ownership (TCO)

Analyzing TCO is vital. TCO takes into account all indirect costs associated with a cybersecurity solution throughout its lifespan. For example, a healthcare provider evaluated TCO to implement a data encryption solution by considering not only the purchase price but also factoring in implementation costs, staff training expenses, and compliance with health data protection regulations.

Assessing the scalability and future expenses of a solution

When it comes to evaluating a solution, it's important to consider its scalability and future costs as this helps us understand how it will fit into our term planning. For example, a growing e-commerce company recently opted for a cloud-based security solution that offered a pay-as-you-grow model. This allowed them to scale their security needs in line with their business growth without having to bear expenses.

Calculating the ROI for cybersecurity investments

Determining the ROI for cybersecurity investments involves comparing the costs of implementing a solution against the benefits it provides. These benefits can range from reducing the risk of breaches and ensuring compliance with regulations to fostering customer trust. To illustrate, let's consider a financial services firm that assessed the ROI of a fraud system. They estimated the losses from fraud that would be prevented by implementing this system while also taking into account the advantages gained through enhanced customer data protection.

Considering indirect benefits and cost savings

It's essential to consider both benefits and cost savings when evaluating cybersecurity solutions. These factors encompass aspects such as increased efficiency, enhanced reputation, and compliance with requirements. For instance, let's look at a manufacturing company that factored in cost savings when deciding on implementing a cybersecurity solution. By doing so, they were able to take into account the reduced downtime and protection against espionage as monetary advantages.

Strategic budgeting for cybersecurity

Budgeting for cybersecurity requires taking an approach that aligns with organizational priorities and risk profiles. A non-profit organization that has a budget gave priority to safeguarding its donor database. They allocated funds accordingly while also looking into cost solutions for critical areas.

Evaluating pricing models

When considering cybersecurity solutions, there are pricing models such as subscription-based options, one-time purchases, or usage-based pricing. A marketing agency conducted a comparison between subscription-based and one-time purchase antivirus solutions to determine which model would provide long-term value for their project-based work environment.

Exploring financial incentives and grants

Certain organizations, non-profits, and educational institutions can benefit from seeking incentives, grants, or subsidized cybersecurity programs. An educational institution took advantage of a government grant program to upgrade its cybersecurity infrastructure. This significantly reduced the burden they faced.

Negotiating with vendors and customization

Engaging in negotiations with vendors can lead to cost savings or the creation of customized solutions. A small retail chain successfully negotiated a tailored package with a security vendor that included features at a reduced cost. This package perfectly aligned with their needs and budget constraints.

Adjusting the cybersecurity budget regularly

It is crucial to review and adjust the cybersecurity budget to keep it in line with the evolving business needs and emerging threat landscapes. As an example, a logistics company conducts evaluations of its cybersecurity expenses by making adjustments to its budget in response to emerging threats, such as an increase in attacks specifically targeting the logistics industry.

Assessing the cost-effectiveness and ROI of cybersecurity solutions is a process that involves understanding the costs, evaluating scalability, calculating ROI, considering indirect benefits, budgeting strategically, exploring various pricing models, seeking financial incentives, negotiating with vendors, and regularly reviewing the cybersecurity budget. By analyzing these financial aspects, organizations can make informed decisions that not only ensure effective protection against threats but also align with their financial capabilities and long-term business objectives. This balanced approach allows organizations to maintain cybersecurity defenses while ensuring financial management.

Implementing and integrating cybersecurity solutions

The crucial stage of implementing and integrating cybersecurity solutions is when strategic plans and theoretical models translate into defenses against cyber threats. In this section, we will guide you through the process of moving cybersecurity solutions from the blueprint phase to their effective real-world application.

Implementing cybersecurity solutions is an undertaking that requires planning, coordination, and execution. It goes beyond installing software or hardware; it involves seamlessly integrating these solutions into your organization's existing technological ecosystem while minimizing disruptions to daily operations. For example, let's consider a scenario where a large financial institution is implementing a network security system. This project would require planning to ensure compatibility with existing software and minimal downtime, which is crucial for maintaining uninterrupted customer service.

Equally important is the aspect of training and user acceptance. Even advanced cybersecurity solutions can fall short if the end users are not trained. In this case, employees are not adequately trained or they do not fully embrace the systems. It is essential to have training programs tailored to user groups within the organization. Imagine a healthcare provider introducing a data encryption tool; they would need training sessions for their staff members.

For the tool to be truly effective, it is crucial that everyone within the organization, from IT professionals to staff, fully embraces it. This requires training programs tailored to accommodate varying levels of expertise.

Another important aspect is the process of monitoring, maintaining, and regularly updating the tool. Cybersecurity cannot be treated as a *set-it-and-forget-it* solution; it demands vigilance and adaptability. The landscape and threat vectors are constantly evolving, necessitating reviews, updates, and maintenance of cybersecurity measures. This may involve conducting security audits implementing software updates and swiftly responding to emerging threats.

Lastly, we'll explore the significance of establishing a feedback loop mechanism for improvement. Cybersecurity implementations should be dynamic and adaptable over time; they are based on user feedback, evolving threats in the realm, and technological advancements. A designed feedback system can result in enhancements in security measures by making them more effective and user-friendly.

Developing a strategic implementation plan for cybersecurity solutions

Implementing cybersecurity solutions within an organization is a vital undertaking that requires strategic planning. A crafted implementation plan not only ensures a smooth process but also guarantees that the chosen solutions align with the organization's specific needs and goals.

Understanding the unique requirements of the organization

Before creating the implementation plan, it is essential to have an understanding of the organization's requirements. This understanding is achieved by analyzing factors such as the structure, workflow, existing systems, and potential cybersecurity risks. For instance, for an institution, protecting customer data and complying with financial regulations are crucial considerations in their plan.

Establishing clear goals and objectives

The implementation plan must outline attainable goals and objectives. These goals could include enhancing data security measures, improving threat detection and response capabilities, or ensuring compliance with regulations. Take, for example, a business aiming to implement a **point of sale (POS)** system equipped with advanced security features to prevent data breaches and fraudulent activities.

Designing the roadmap for implementation

A roadmap plays a role in successful implementation. It should encompass milestones, timelines for completion, resource allocation details, as well as roles and responsibilities. In the case of implementing an electronic health records system within a healthcare organization, an effective approach might involve phased implementation with defined milestones for each department.

Coordination and communication

Effective coordination and clear communication among all parties involved are crucial throughout the implementation process. Regular meetings, updates, and collaborative tools can greatly facilitate this. For instance, when a multinational company introduced a network security solution, it utilized a project management tool to keep all teams aligned and well-informed.

Addressing technical and operational challenges

It is important for the plan to address any operational obstacles. This includes ensuring compatibility with existing systems, seamless data migration, and minimizing disruptions to operations. To illustrate, a manufacturing company might strategize implementing the solution during off-peak hours to minimize its impact on production.

Training and support

Provisioning training sessions and ongoing support is critical for the adoption of new cybersecurity solutions. This involves training as well as continuous assistance. As an example, an educational institution that implemented security software organized training sessions for its staff members while also establishing a dedicated support team to provide ongoing help.

Testing and validation

Conducting thorough testing and validation of cybersecurity solutions before full-scale deployment is essential. This can involve conducting pilot tests in controlled environments. For instance, a logistics firm carried out a test of an intrusion detection system in one of their warehouses before implementing it across the entire company.

Risk management and contingency planning

A contingency plan should include strategies to manage risks and prepare for issues that may arise during the implementation phase. For example, a technology start-up could anticipate delays or compatibility problems with existing software.

It is important to incorporate monitoring and feedback mechanisms into the plan. This allows you to track the progress of implementation and make adjustments. To illustrate, a service company established a feedback loop with its IT department to monitor the performance of implemented cybersecurity measures and make adjustments.

Developing an implementation plan for cybersecurity solutions is a process. It involves considering an organization's needs, setting clear objectives, creating a detailed roadmap, ensuring open communication and coordination, addressing technical challenges, prioritizing training, conducting thorough testing, managing risks effectively, and establishing monitoring and feedback mechanisms. By following these steps, organizations can successfully implement cybersecurity solutions that align with their needs and objectives.

User training and adoption in cybersecurity implementation

Successfully implementing cybersecurity solutions not only relies on just the deployment of the technology but also greatly depends on user training and adoption. It is crucial to have a designed training program that ensures all users understand how to use the new systems and recognize their role in maintaining cybersecurity.

To start, it is important to assess the training needs of user groups within your organization. Different roles may require varying levels of training. For example, IT staff will need training, while other employees may only need to familiarize themselves with specific features or be able to identify phishing attempts.

Next, design a tailored training program that caters to the needs of each user group. For instance, an international bank that implements encryption software developed training modules for different departments, ensuring that each team understood the relevant aspects of the software.

Choose training methods that align with your organization's culture and match the learning styles of your employees. Interactive workshops, webinars, and eLearning modules are some examples of methods. To illustrate this point further, a technology company utilized a combination of in-person workshops and interactive eLearning courses to train their staff on an implemented security protocol.

It is also beneficial to involve management in the training process as their participation helps emphasize the importance of cybersecurity throughout the organization.

When a healthcare provider implemented a system to safeguard data involving management, in-training sessions highlighted the company's dedication to cybersecurity throughout the organization.

Ongoing refresher courses

The field of cybersecurity is constantly changing, so it's crucial to engage in learning. Regularly providing refresher courses and updates on emerging threats helps ensure that knowledge remains up to date. For instance, an e-commerce company introduced cybersecurity refreshers for its staff, informing them about the latest security risks and best practices.

Addressing resistance to change

Resistance toward systems is common. It can be overcome by addressing concerns during the training process. Understanding employee's apprehensions and actively working toward alleviating them can facilitate a transition. For example, a manufacturing firm initially faced resistance when implementing an intrusion detection system; however, they successfully addressed this by communicating the benefits and offering comprehensive hands-on training.

Evaluating training effectiveness

It is essential to monitor and evaluate the effectiveness of your training program. This can be achieved through assessments, feedback forms, or tracking cybersecurity incidents related to user errors. A retail chain implemented cyber drills and assessments as part of its cybersecurity training program to assess its efficacy.

Continuous improvement through a feedback loop

Establishing a feedback loop improves the training program. By seeking feedback from participants, adjustments can be made to enhance iterations of the training sessions.

To successfully implement cybersecurity solutions, it is important to consider training and user adoption. Here's how organizations can achieve this:

- **Assess training needs**: Gather employee feedback to understand their training requirements and use this information to make adjustments. For example, a software development company regularly gathered feedback after training sessions to improve its cybersecurity curriculum.

- **Tailored training programs**: Design customized training programs that address the needs of employees. This ensures that the training is relevant and engaging for everyone involved.

- **Effective training methods**: Implement effective training methods that cater to learning styles. This could include workshops, online modules, or hands-on simulations.

- **Management involvement**: Engage management in the training process by emphasizing their role in cybersecurity initiatives. This helps create a culture of accountability and encourages employee participation.

- **Continual learning**: Encourage learning by providing resources such as support desks, FAQs, and resource libraries dedicated to cybersecurity topics. A logistics company even set up a support desk for employees seeking assistance with security systems.

- **Overcome resistance**: Address any resistance or skepticism toward cybersecurity measures through communication about the importance of protecting data and the potential consequences of breaches.

- **Monitor effectiveness**: Regularly evaluate the effectiveness of the training programs by tracking metrics such as employee knowledge retention or incident rates related to cyber threats.

- **Establish a feedback loop**: Maintain a feedback loop where employees can share their experiences with the training programs and suggest improvements if needed.

- **Support adoption**: Provide support resources such as refresher courses or updated materials to ensure employees' continuous adoption of security practices.

By following these steps, organizations can build a cybersecurity culture within their workforce while ensuring all employees effectively use and embrace these security measures.

Monitoring, maintaining, and regularly updating cybersecurity solutions

Maintaining cybersecurity requires effort rather than a one-time event. It is crucial to monitor, maintain, and regularly update security measures to stay ahead of evolving threats.

To begin with, it is important to establish systems for monitoring. These systems should have the ability to detect threats, breaches, and vulnerabilities in time. For instance, a financial institution may implement advanced network monitoring tools to scrutinize data for any signs of fraudulent activity.

Routine maintenance plays a vital role in ensuring the effectiveness of cybersecurity systems. This involves updating security protocols, patching vulnerabilities, and checking system integrations. For example, a retail company may schedule maintenance checks to keep its POS systems up to date and ensure that all security measures are functioning correctly.

Regular updates are necessary to protect against cybersecurity threats. This includes updating software, reinforcing security protocols, and staying informed about the developments in cybersecurity. A healthcare provider could establish a policy for updates on their security software to consistently safeguard patient data.

Involving employees in the monitoring process can greatly enhance your cybersecurity posture. Educating them about how to recognize and report activities adds a layer of security.

A technology company, for instance, trained its employees on how to identify and report phishing attempts, which greatly reduced the risk of email-based attacks.

Using automated tools to improve efficiency

Automated tools can significantly enhance the efficiency and effectiveness of cybersecurity monitoring. For example, AI-powered threat detection systems can provide real-time alerts and lighten the workload on IT staff. An e-commerce business could employ automated scanning tools to continuously monitor their website for any security breaches.

Carrying out security audits

Regular security audits are crucial in evaluating the state of your cybersecurity infrastructure. These audits should be comprehensive covering all aspects of your cybersecurity measures. An educational institution might conduct biannual security audits to assess the effectiveness of its cybersecurity measures and identify areas for improvement.

Responding to security incidents

Having a defined protocol for handling security incidents is a part of maintaining effective cybersecurity. This includes steps such as containment, eradication, recovery, and post-incident analysis. For instance, a logistics company swiftly contained an attack and successfully recovered its data by following carefully planned incident response protocols.

The importance of third-party services

In some cases, utilizing third-party services for monitoring and maintenance can be advantageous for organizations with internal IT resources. For instance, a small business could consider collaborating with a cybersecurity company to receive monitoring and maintenance services.

Continuous improvement through feedback and analysis

Cybersecurity is an evolving field and it's important to continuously improve our security measures. Gathering feedback from system users and analyzing security incidents can offer insights that help enhance our security protocols. For instance, a manufacturing company implemented a feedback system for their IT team, which resulted in improvements in their network security.

To ensure cybersecurity, it's crucial to incorporate monitoring, maintenance, and regular updates into our strategy. This involves implementing monitoring systems, performing maintenance tasks, involving employees in the process, leveraging automated tools whenever possible, conducting periodic security audits, preparing for incident responses, considering third-party services when necessary, and maintaining a focus on continuous improvement. By following these practices, organizations can ensure that their cybersecurity infrastructure remains strong and adaptable to threats while countering current ones.

Summary

In this chapter, we gained insights into the process of choosing, implementing, and integrating cybersecurity solutions that work effectively within an organization.

We began by exploring the factors to consider when selecting cybersecurity solutions. This involved understanding the threats that are relevant to your industry and organization, evaluating how well potential solutions align with your existing IT infrastructure, and considering their scalability to accommodate future growth. We also emphasized the importance of adhering to requirements and industry standards, underscoring the need for cybersecurity measures that are both effective and compliant with legal obligations.

We then moved on to discussing practices for selecting cybersecurity solutions. We highlighted conducting market research to explore a range of available options and their features. Additionally, we stressed involving stakeholders in the selection process to ensure that chosen solutions meet the needs and objectives of various departments in your organization. Furthermore, we delved into risk assessment and management techniques for prioritizing solutions based on their ability to mitigate risks effectively while also evaluating cost-effectiveness and ROI.

Finally, this chapter delved into various aspects of implementing and integrating cybersecurity solutions. It emphasized the importance of creating a thought-out implementation plan to ensure an efficient deployment of these measures. Additionally, this chapter highlighted the role played by training and user adoption in enabling employees to utilize these solutions. The significance of monitoring, maintenance, and regular updates was also stressed as cybersecurity is a process that necessitates constant attention. Lastly, this chapter underscored the value of establishing a feedback loop and a mechanism for improvement, allowing organizations to adapt and enhance their cybersecurity measures over time.

To summarize, this chapter offered a guide on how to select, implement, and integrate cybersecurity solutions. It covered everything from understanding selection factors to following practices during the selection process. Implementing these solutions while prioritizing improvement organizations can significantly enhance their cybersecurity posture. Armed with this knowledge, you are now better equipped to make decisions that align with your organization's cybersecurity needs and objectives.

In the next chapter, you will learn how to bridge the gap between technical and non-technical stakeholders, enabling you to effectively engage with all stakeholders across your organization.

10
Bridging the Gap between Technical and Non-Technical Stakeholders

As cybersecurity threats become more advanced, all departments must communicate clearly, work together cohesively, and collaborate effectively. This chapter emphasizes that cybersecurity has become a board-level matter and requires effective collaboration between technical and non-technical stakeholders. The ability to convert information into actionable insights for business purposes is valuable. The goal is to simplify jargon and align perspectives toward a shared objective. It goes beyond sharing information; it involves transforming how that information is perceived and used across layers of an organization.

We will explore why effective communication and collaboration are critically important in cybersecurity. This exploration will focus on understanding the "what" and "why" and also delve into strategies for creating an environment where technical and non-technical individuals can thrive together to address cybersecurity challenges. Subsequently, we will examine how to translate concepts for those who may not have a strong technical background.

This skill goes beyond language proficiency; it requires empathy, comprehension, and the ability to bridge the gap between risks and their impact on business. It's about creating a connection that breaks down the language barrier separating experts from those who aren't as technically inclined.

In addition, we will concentrate on strategies to foster collaboration. This surpasses conversation; it entails establishing integrated workflows, shared goals, and mutual respect between non-technical teams. By cultivating a culture that appreciates and comprehends the contributions of each stakeholder, organizations can come together to combat cybersecurity threats.

This chapter serves as a guide for both technical experts looking to enhance their communication skills and non-technical professionals aiming to grasp the intricacies of cybersecurity. It's a journey toward establishing a language of shared understanding and a collaborative approach to safeguarding our world.

We'll cover the following topics in this chapter:

- The importance of effective communication and collaboration
- Translating technical concepts for non-technical stakeholders
- Strategies for successful collaboration between technical and non-technical stakeholders

The Importance of Effective Communication and Collaboration

So, let's start exploring how different individuals involved in cybersecurity interact with each other. We must start by acknowledging and understanding the obstacles in communication between experts and those not technically inclined. These obstacles, which range from language to differing priorities, can significantly impede the effectiveness of cybersecurity strategies.

Understanding communication barriers in cybersecurity

Effective communication plays a role in cybersecurity. However, one significant challenge remains – the communication gap between technical and non-technical individuals. This gap often arises due to the terminology and concepts associated with cybersecurity, leading to misunderstandings, misaligned priorities, and weakening security measures. Such terms include "zero-day vulnerability" and "endpoint encryption." While these terms may be commonplace for cybersecurity professionals, they can be completely unfamiliar to technical executives. As a result, there may be a lack of understanding regarding the urgency and importance of security measures.

A key aspect of this barrier relates to the perspectives regarding priorities. While technical professionals emphasize the significance of "encryption protocols" and "firewall integrity," non-technical stakeholders are more concerned about how these aspects affect data privacy laws, customer trust, and regulatory compliance. This difference in focus can lead to strategies unless there is an effort to translate priorities into their business-related consequences.

The impact of these communication barriers goes beyond miscommunication; it can influence the cybersecurity culture within an organization. When terms such as "penetration testing" or "risk assessment" are not adequately explained, non-technical stakeholders may not fully comprehend why certain investments in cybersecurity are necessary, resulting in a lack of prioritization for security measures.

To bridge this gap, creating an environment where technical terms are demystified and cybersecurity is presented in terms of business impact is crucial. Upon mentioning "SSL/TLS encryption," a technical expert could explain it as a method to safeguard customer data and maintain trust, thus aligning it with business goals. Establishing this language requires patience, empathy, and a comprehensive understanding of both the technical details and the business implications of cybersecurity.

The communication obstacles within the cybersecurity field, primarily caused by language, pose a considerable challenge to effective security management. To overcome these barriers, it is crucial to make a conscious and thoughtful endeavor to harmonize the viewpoints and priorities of all those involved. By promoting a culture that encourages accessible and understanding communication, organizations can guarantee that their cybersecurity strategies are not only technically robust but also aligned with their broader business goals.

The role of effective communication in cybersecurity success

Let's shift the focus from understanding barriers to communication to exploring how effective communication actively contributes to the success of cybersecurity efforts. In this context, effective communication goes beyond transmitting information; it involves ensuring that the information is received, understood, and acted upon in a way that aligns with the organization's cybersecurity objectives. Let's look at some of the key aspects of communication in cybersecurity:

- One crucial aspect of communication in cybersecurity is **conveying risk**. For instance, when discussing a security vulnerability, it is essential to explain the level of risk in a way that all stakeholders can comprehend. While a technical team may understand the implications of a SQL injection flaw for technical individuals, describing it in terms of possible data breaches and legal consequences makes the risk more tangible. This translation is vital for enabling technical stakeholders to make informed resource allocation and risk management decisions.

- Another integral component is **communicating cybersecurity policies and procedures**. Often, these are documented using language that may be inaccessible to employees who are not cybersecurity experts. For example, when it comes to secure password management, it's important to communicate the policy in a way that it's both clear and easy to understand. This helps emphasize the significance of protecting company data. Following the policy goes beyond rules; it becomes a shared responsibility and an integral part of the organizational culture.

- **Effective communication** also plays a role in incident response. During a cybersecurity incident, timely and transparent communication is crucial. This includes the response team and other stakeholders, such as public relations, legal departments, and upper management. For instance, during a data breach, while the technical team focuses on containment and mitigation efforts, it's essential to keep stakeholders informed about the breach's nature, potential impact, and steps being taken to address it. This coordinated approach ensures an efficient response.

- **Training and awareness programs** are also areas where effective communication is key. Cybersecurity training should be a process that adapts to evolving threats rather than just a one-time event. For instance, when introducing phishing tactics, it's important to explain why this training matters, how it relates to employee's daily work routines, and what steps they can take to identify, and then report suspected phishing attempts.

The importance of communication in cybersecurity cannot be emphasized enough. It ensures that all individuals involved, regardless of their knowledge, understand cybersecurity risks, policies, incident response procedures, and training. Organizations prioritizing clear, relatable, and practical communication methods can greatly improve their cybersecurity stance and integrate it seamlessly into their operations.

Strategies for successful collaboration between technical and non-technical stakeholders

Focusing on methods to bridge the gap between distinct groups within an organization is critical for our joint success in improving the organization's security posture. It emphasizes how collaboration can be improved through approaches and techniques, ensuring that all parties contribute effectively to the organization's cybersecurity measures.

One important approach is **creating functional teams** that include members from technical and non-technical backgrounds such as IT, legal, human resources, and business operations. For example, when implementing a security protocol, a cross-functional team can ensure that the protocol meets both security standards and aligns with legal requirements and operational workflows. This fosters an understanding of cybersecurity that considers all aspects of the organization.

Another key strategy is **maintaining structured communication**. This can be achieved through scheduled meetings, updates, and collaborative platforms where information is shared and discussed. For instance, conducting cybersecurity briefings where IT leaders present updates and initiatives using accessible language can encourage better understanding and engagement across the organization in terms of non-technical staff.

Creating shared goals and objectives is also crucial. These objectives should be clearly defined, measurable, and in line with both the mission of the organization and specific cybersecurity needs. For instance, a common goal could be to decrease susceptibility to phishing attacks by a percentage. By having a shared objective, technical and non-technical stakeholders can collaborate effectively toward an endpoint.

Developing a shared language is another aspect. This involves establishing a set of terms and concepts that everyone in the organization can understand because of their background. Instead of using terms such as "phishing," for example, the team can use more descriptive language, such as "fraudulent emails that attempt to deceive employees."

Involving technical stakeholders in decision-making processes related to cybersecurity is also essential. This inclusion ensures that decisions reflect perspectives and that their implications are understood throughout the organization. For instance, when selecting a cybersecurity solution, it is invaluable to receive input from the finance department regarding budget constraints and from the team regarding compliance requirements.

Mutual education and training are also strategies to consider. Technical teams can provide updates on the state of cybersecurity in an easily understandable manner, while non-technical teams can educate IT professionals about business processes and customer viewpoints.

For instance, hosting a workshop where the marketing team elucidates their utilization of customer data can aid IT professionals in comprehending the significance of safeguarding data.

Lastly, **acknowledging and celebrating collaboration** can reinforce its worth. Recognizing the accomplishments of functional teams in achieving cybersecurity objectives through company-wide communications or awards can inspire ongoing collaborative endeavors.

In conclusion, effective collaboration between non-technical stakeholders in cybersecurity is not only desirable but crucial. By employing strategies such as functional teams, regular communication, shared objectives, a common language, inclusive decision-making processes, mutual education, and acknowledgment of collaborative efforts, organizations can establish a more unified, knowledgeable, and efficient approach to managing cybersecurity risks. These strategies ensure that cybersecurity is ingrained in every aspect of the organization with ongoing effort from all its members.

Translating technical concepts for non-technical stakeholders

Let's delve into a crucial aspect of cybersecurity communication: the ability to translate complex technical ideas into clear and relevant business language. This translation process goes beyond simplifying terminology; it aims to make technical information accessible and meaningful to individuals without a background. The ultimate goal is to foster a comprehension and appreciation of cybersecurity matters at all levels within an organization.

We'll commence by discussing the art of *simplifying complex cybersecurity terminology*. In this section, we will focus on the challenge of breaking down jargon into terms without diluting the essence of the information. This skill is particularly vital in cybersecurity, where topics such as network security, encryption, and malware can be overwhelming for technical stakeholders. The ability to simplify these concepts while maintaining their integrity is essential for communication.

Moving forward to *Contextualizing cybersecurity in a business context*, we will explore how cybersecurity professionals can present issues within a business framework. This involves translating the consequences of cybersecurity threats and measures into their impact on business operations, finances, compliance, and customer relations. By doing this, technical teams can effectively communicate the significance of investing in cybersecurity initiatives to decision-makers.

Finally, in the *Effective visualization and presentation of cybersecurity data* section, we will explore the importance of communication. Our focus here is on transforming sets of data, such as threat analysis reports or cybersecurity metrics, into formats that are easily comprehensible. By utilizing charts, graphs, and infographics, we can effectively present information in a manner that engages and informs technical audiences.

This particular section aims to equip cybersecurity professionals with the skills to bridge the communication gap between themselves and non-technical stakeholders. By mastering the art of translating concepts into business language, cybersecurity professionals can ensure that their expertise and insights are not only heard but also understood and valued within their organizations. It is not about communicating; it is about fostering shared understanding and adopting an approach to managing cybersecurity risks.

Simplifying complex cybersecurity terminology

A key step in simplifying cybersecurity terminology is to use analogies and metaphors that relate to experiences. For example, if we are trying to explain what a "firewall" is on its own, you can compare it to a security guard stationed at the entrance of a building. This security guard controls who enters and exits the building, making the concept easier to grasp. Similarly, you can compare "malware" to a flu virus infiltrating our bodies, with computers or networks being our bodies in this analogy. These comparisons demystify terms and also help to create mental images that non-technical individuals can easily remember.

Another effective approach is to **replace terms with language explanations**. For instance, when using the term "phishing," you could describe it as "a method where attackers trick people into revealing information such as passwords through deceptive emails." This technique involves breaking down processes into elements that are easy for anyone to understand.

Storytelling can be a tool to simplify jargon. By using narratives that involve real-life scenarios or hypothetical situations, we can illustrate how cybersecurity concepts apply more comprehensively. For example, sharing a story about how a simple password reset prevented a data breach can effectively emphasize the importance of cybersecurity practices such as regularly updating passwords.

Visual aids play a role in demystifying terms. Diagrams, flowcharts, and infographics are resources to visually represent concepts such as "encryption" or "network security," making them easier to comprehend. For instance, employing a flowchart that explains how encryption works to secure data can help non-technical individuals visualize the process and grasp its significance without getting overwhelmed by details.

Incorporating elements such as simulations or basic cybersecurity games can also be highly effective in simplifying and explaining complex ideas. These interactive elements allow stakeholders to engage hands-on and experience firsthand the importance of cybersecurity practices. For instance, through a simulation demonstrating a network under fire from a **distributed denial-of-service (DDoS)** attack, participants gain insights into the impact such attacks have on business operations.

When it comes to making complex cybersecurity terms easier to understand, it takes a mix of creativity, empathy, and a solid grasp of both the aspects and how the audience sees things. By using analogies, language, storytelling techniques, visuals, and interactive elements, cybersecurity experts can effectively convey ideas in a way that everyone can grasp them and find them interesting. This method not only improves comprehension but also promotes a more inclusive and well-informed approach to handling cybersecurity risks within an organization.

Contextualizing cybersecurity in business terms

In this section, we'll learn how to explain cybersecurity in a way that makes sense to technical stakeholders. This essential aspect of communication involves translating cybersecurity concepts into the context of business operations and goals.

To truly grasp the impact of cybersecurity on businesses, it is crucial to understand its implications on financials. For example, a data breach can lead to losses due to fines or costs associated with remediation and decreased revenue. By providing examples such as the cost per lost record or the effect on share prices, we can help non-technical stakeholders understand why robust cybersecurity measures are financially urgent.

Another important aspect is compliance and legal considerations. Cybersecurity goes beyond safeguarding information; it also involves adhering to requirements. By illustrating how non-compliance with regulations such as the **General Data Protection Regulation (GDPR)** can result in fines and legal complications, we make the topic relevant not for departments, such as legal and compliance, but for executive-level decision-makers.

Furthermore, an organization's cybersecurity practices significantly impact customer trust and brand reputation. For instance, when a company experiences a high-profile cyberattack, it can encounter challenges such as losing customers and damaging its brand reputation. This can have long-term consequences for revenue. Making a connection between cybersecurity practices and customer retention rates is important for departments focused on customer experience and marketing.

Business continuity is also heavily influenced by cybersecurity. If a ransomware attack occurs and prevents access to data, it can disrupt operations and result in immediate financial losses, as well as harm relationships with suppliers and customers in the long run. Explaining the significance of cybersecurity within business continuity plans and operational resilience underscores its importance for teams and those involved in planning.

Moreover, cybersecurity plays a role in fostering business innovation. As businesses increasingly rely on technologies, establishing a foundation is vital for exploring new markets or launching innovative products. For example, showcasing how a strong cybersecurity framework enables entry into e-commerce aligns IT objectives with business development strategies.

Considering investor relations is also essential. Investors now view cybersecurity as an indicator of a company's stability and long-term viability. Describing cybersecurity initiatives in terms of risk management and corporate governance makes the topic relevant to stakeholders involved in investor relations and corporate strategy.

When we talk about cybersecurity concerning businesses, it is about understanding and explaining the implications of technical risks and actions in a broader business context. This means not only highlighting how cybersecurity directly affects areas of business operations but also ensuring that cybersecurity initiatives are aligned with the organization's overall strategic objectives. By doing this, we can integrate cybersecurity into business discussions and make sure that its significance is acknowledged and given priority at every level of the organization.

Effective visualization and presentation of cybersecurity data

When it comes to visualization, using **graphs and charts** is key in representing cybersecurity metrics. For instance, a line graph that shows the number of phishing attempts detected over time can visually demonstrate the increasing or decreasing risk, making it easier for non-technical stakeholders to grasp the severity and frequency of threats. Similarly, a pie chart can effectively depict the proportion of types of malware attacks faced by an organization.

Infographics are also tools for simplifying cybersecurity data. They combine text, images, and data to convey a story or explain concepts in an easy-to-understand format. For example, an infographic explaining an attack could illustrate the sequence of events from the breach to the recovery process while highlighting important steps and decisions made along the way.

Dashboards provide a means of presenting real-time data. A designed cybersecurity dashboard can offer a snapshot of an organization's security status by displaying indicators such as threat levels, recent incidents, and compliance status.

This allows people without backgrounds to quickly understand the situation without having to get into the details.

Interactive data visualizations take engagement to the next level. Interactive features, such as maps that display cyber threats, enable individuals to explore data more deeply based on their interests and needs. This kind of interactivity not only makes the data more captivating but also allows users to personalize their experience and gain specific insights that are relevant to their roles or concerns.

Visualizing cybersecurity incidents through scenarios can also be highly effective. Creating representations of scenarios, such as a data breach or system compromise, helps non-technical stakeholders grasp the implications of threats and understand the importance of various security measures. This approach is particularly valuable in training and awareness sessions.

Comparative visuals are valuable for showcasing the effectiveness of cybersecurity investments. For instance, using before and after visualizations that demonstrate a reduction in attacks or security incidents after implementing a security tool can highlight the return on investment for management and financial stakeholders.

Telling stories through data visualization is another technique we can use. By presenting data in a narrative format, complex cybersecurity concepts can be made relatable and memorable. Showing the evolution of cybersecurity threats over the decade through a timeline graphic can provide context and help us understand how the landscape has changed.

When it comes to data visualization, simplicity is crucial. Complicated charts or graphs can be as confusing as jargon. The main objective is to simplify and focus on conveying the core message in the best way possible.

Effectively visualizing and presenting cybersecurity data goes beyond converting numbers into pictures. It is about creating a story that makes complex data accessible, engaging, and meaningful. By using graphs, infographics, dashboards, interactive visualizations, and storytelling techniques, cybersecurity professionals can effectively communicate information to technical stakeholders. This fosters an understanding and appreciation of cybersecurity challenges and their solutions.

Now, let's shift our attention to implementing collaboration strategies between technical and non-technical individuals in the field of cybersecurity. The next section focuses on providing methods and approaches to cultivate a cohesive working relationship among different teams within an organization.

Strategies for successful collaboration

Collaboration goes beyond bringing people. It involves leveraging the diverse strengths and perspectives that each team member brings to the table to enhance the organization's cybersecurity stance.

We'll commence with *building cross-functional cybersecurity teams*. In this section, we'll emphasize forming teams that combine expertise with business acumen. The objective is to create groups where members complement one another's skills, enabling an approach to address cybersecurity challenges. This section will delve into practices for team formation, highlighting the significance of diversity in terms of skills and backgrounds, as well as tactics to ensure effective teamwork.

Moving on to *Establishing regular cybersecurity workshops and training sessions*, we will address the necessity for education and shared learning experiences. These sessions play a role in keeping both non-technical stakeholders well-informed and engaged. They offer a platform for learning and a space for discussions where insights from various business areas can contribute to a more comprehensive and well-informed approach to cybersecurity.

Lastly, in the *Implementing collaborative cybersecurity decision-making processes* section, we will delve into how organizations make decisions concerning cybersecurity. This part will explore how to establish a decision-making framework that listens to inputs from different stakeholders and actively incorporates them. This involves creating structures and procedures that encourage discussions, thereby enabling informed decisions that align with technical requirements as well as business objectives.

This particular section is crucial as it brings together the concepts we've discussed – communication and translating technical ideas – and applies them in practical collaborative settings. The goal is to equip cybersecurity professionals with the knowledge and tools to cultivate an environment where collaboration becomes a reality rather than just an aspiration, thereby effectively bolstering the cybersecurity resilience of their organizations.

Building cross-functional cybersecurity teams

The initial step in establishing a functional cybersecurity team involves identifying the appropriate blend of skills and expertise. This entails selecting team members who possess not only knowledge of cybersecurity but also backgrounds in business operations, legal compliance, human resources, and other relevant areas. For example, including someone from the legal department can provide insights into compliance matters while a marketing professional can offer their perspective on safeguarding customer data and maintaining brand reputation.

Once the team has been assembled, one of the challenges is **fostering communication**. This entails creating a shared language that goes beyond jargon so that all members can contribute meaningfully. Conducting team-building activities and workshops can greatly assist in breaking down barriers and cultivating understanding, as well as respect.

Additionally, it is essential to **establish roles and responsibilities** within the team. Ensuring clarity in roles is crucial to ensure that every team member understands their contribution toward achieving the team's objectives. For instance, IT professionals may be responsible for implementing security measures while representatives from the business side can play a role in evaluating the potential business impacts of those measures.

It is essential to **establish shared goals and objectives** as it helps align the efforts of the team. These goals should encompass both the aspects of cybersecurity and the broader implications for the business. For example, a shared goal could be to reduce the organization's cyber risk exposure by a percentage within a year.

Creating a culture of collaboration is vital for success. This can be accomplished by providing opportunities for team members to collaborate on projects, share insights, and learn from one another. Regular meetings that allow members to discuss issues, brainstorm solutions, and share updates can greatly contribute to fostering this culture.

Measuring the effectiveness of the team is another aspect to consider. Establishing metrics and **key performance indicators (KPIs)** related to cybersecurity while also reflecting business outcomes can help evaluate the team's performance. For instance, tracking improvements in incident response time or measuring enhancements in employee awareness regarding cybersecurity can provide metrics.

Leadership plays a significant role in guiding and supporting functional teams. Leaders should play a role in not only giving guidance and resources but also in actively fostering an inclusive and collaborative culture. They need to champion the idea of functional teamwork, highlighting its achievements and learning from the obstacles it presents.

It is important to have mechanisms in place for resolving conflicts. When you have a range of perspectives, disagreements are bound to occur. By establishing predefined processes for conflict resolution, the team can effectively navigate these challenges and maintain interactions.

The development of functional cybersecurity teams requires meticulous planning, effective communication, and a strong commitment to collaboration. By bringing individuals with different perspectives and skill sets, organizations can create teams that are well-prepared to tackle the complex nature of cybersecurity challenges. These teams serve as a foundation for a proactive cybersecurity strategy that does not only respond to threats but also anticipates them and effectively mitigates their impact.

Establishing regular cybersecurity workshops and training sessions

Shared learning opportunities can help bridge the gap between technical and non-technical individuals. These workshops and training sessions play a role in establishing an understanding of cybersecurity principles, risks, and best practices within an organization.

Implementing workshops and training sessions begins by identifying the requirements of distinct groups within the organization. Conducting a needs assessment helps identify areas where knowledge gaps exist. For instance, non-technical staff may require cybersecurity awareness training, while technical staff may benefit from updates on threat landscapes or advancements in security technologies.

Creating relevant content for all participants is crucial. This involves developing training materials that are easily understandable for technical audiences but still valuable for technical staff members. Real-world scenarios or case studies can be used to make the content relatable and practical. Additionally, interactive elements, such as hands-on activities or simulations, can enhance engagement and facilitate learning experiences.

When scheduling these sessions, it is important to consider business operations to minimize disruption. Regular shorter sessions may be more effective than infrequent ones.

For example, hosting a workshop lasting 1 hour can be more manageable and less disruptive compared to a full-day seminar held once a year.

It is important to consider learning styles when delivering training. Some team members may prefer presentations while others may benefit more from hands-on exercises or group discussions. By offering a variety of teaching methods, such as lectures, workshops, and interactive activities, we can ensure that everyone has a learning experience.

Bringing in experts or guest speakers can bring perspectives and insights. For instance, inviting a cybersecurity consultant to discuss the trends or having a legal expert talk about compliance issues can enrich the learning process by providing real-world knowledge.

To continuously improve, it's crucial to have feedback mechanisms in place. Conducting session surveys or discussions allows us to gather feedback on the effectiveness of the training so that we can make necessary adjustments and enhancements for future sessions. This feedback is also valuable in identifying areas that require coverage or different teaching approaches.

Fostering a culture of learning is essential. Apart from training sessions encouraging self-learning through resources, newsletters, and informal study groups enable individuals to constantly develop and build upon the knowledge that's gained during workshops.

Measuring the impact of these sessions on cybersecurity awareness and practices holds significance. There are ways to measure the effectiveness of training programs, such as tracking changes in incident rates improvements in compliance, or receiving feedback from employees. These metrics provide evidence of how the training has been implemented.

Cybersecurity workshops and training sessions play a role in fostering a knowledgeable organizational culture. It is essential to customize these sessions to cater to the needs of the workforce and utilize teaching methods. Additionally, continuous evaluation of their effectiveness is necessary for organizations to enhance their cybersecurity posture. These educational initiatives raise awareness and empower all members of the organization to actively contribute toward a more secure and resilient cyber environment.

Implementing collaborative cybersecurity decision-making processes

Now, let's explore how to integrate perspectives and inputs into the framework to be able to make cybersecurity decisions.

The first step in establishing a decision-making process is to **create a group of decision-makers**. This group should include representatives from departments such as IT, legal, finance, operations, and human resources. For example, when discussing the adoption of security technology, the IT team can provide insights and apply technical and operational evaluation criteria, while the finance team can evaluate budget implications and the legal team can consider compliance requirements.

It is crucial to **establish channels of communication** to facilitate information sharing and exchange of perspectives. This could involve meetings, shared platforms, or dedicated communication lines. For instance, an online forum or dashboard where upcoming decisions are posted and feedback is encouraged can promote participation from stakeholders.

Defining the decision-making process itself is also particularly important. This means outlining the steps involved in deciding – from proposal and consultation, all the way to approval. For instance, when introducing a cybersecurity policy, the process usually begins with a proposal from the IT department. Feedback is sought from departments. Eventually, it gets finalized by the executive team.

It is important to **incorporate feedback mechanisms** throughout this process. This allows stakeholders to express their concerns, provide suggestions, and show support for decisions. Tools such as surveys, focus groups, or feedback sessions can effectively gather this input.

To resolve conflicts and bring together perspectives, it is beneficial to employ **consensus-building techniques**. Methods such as the Delphi approach, where anonymous feedback is collected and summarized until agreement is reached, or facilitated workshops can help establish understanding and consensus.

Maintaining transparency during the decision-making process plays a huge role. All stakeholders should be kept informed about decision progress, the reasoning behind them, and their expected impact. This openness fosters and ensures alignment among all involved in decision-making.

Providing training and education on the decision-making process can further enhance participation. By understanding how decisions are made, the criteria considered, and their impact, stakeholders can contribute effectively.

To gain insights before the full-scale implementation of decisions, it can be beneficial to carry out **pilot programs or trials**. For instance, let's consider the scenario where a new security protocol is about to be implemented in the organization. Before going with this implementation, it would be wise to conduct a pilot in one department. This approach allows us to identify any challenges or areas that need improvement.

It is crucial to review and evaluate the decision-making process to ensure its effectiveness and adaptability. This involves analyzing the outcomes of decisions, gauging stakeholder satisfaction, and assessing the efficiency of the process. By doing this, we can make improvements.

Implementing a collaborative cybersecurity decision-making process is both intricate and indispensable. By involving a range of stakeholders, establishing lines of communication, defining a structured decision-making process, and prioritizing transparency and adaptability, organizations can make well-rounded decisions that effectively tackle cybersecurity challenges while aligning with their organizational goals. Such an inclusive approach enhances decision quality and also fosters a culture of shared responsibility and trust throughout the organization.

Summary

In this chapter, we explored the relationship between non-technical stakeholders in cybersecurity. This chapter was organized into three sections, each with subsections aiming to provide a practical understanding of effective collaboration in cybersecurity.

The first section, *The significance of communication and collaboration*, highlighted the importance of bridging communication gaps within the realm of cybersecurity. It discussed overcoming communication barriers, emphasizing how effective communication plays a role in achieving success in cybersecurity endeavors. Additionally, it emphasized the need to foster a culture within cybersecurity teams. This section covered topics ranging from tackling jargon to aligning cybersecurity objectives with business goals. Doing this lays a foundation for interdisciplinary collaboration.

Moving on to the *Simplifying technical concepts for non-technical stakeholders* section, we delved into the art of making complex cybersecurity topics accessible for everyone. This section provided guidance on simplifying cybersecurity terminology, contextualizing these issues using business terminology, and effectively visualizing cybersecurity data. Equipped with these skills, you will be able to convey information and demonstrate how cybersecurity impacts various aspects of business operations through visual aids.

Finally, in the *Strategies for successful collaboration* section, the focus shifted to ways to improve teamwork among diverse groups. The subsections discussed approaches such as forming functional cybersecurity teams, organizing regular workshops and training sessions, and implementing collaborative decision-making processes. From building teams to creating educational programs and incorporating different perspectives in decision-making, this section presented actionable strategies for promoting a collaborative culture.

This chapter provided a set of tools for cybersecurity professionals, equipping them with the knowledge and skills needed to collaborate effectively with technical stakeholders. By grasping and applying these concepts, professionals can ensure that cybersecurity becomes more than a priority – it becomes a shared responsibility across all levels within an organization.

In the next chapter, we will delve into the building blocks required to raise a cybersecurity-aware organizational culture. Remember that the responsibility of keeping an organization secure lies with each of the employees and not in isolation from one department.

11

Building a Cybersecurity-Aware Organizational Culture

In this chapter, we will delve into the importance of cultivating a cybersecurity mindset at all levels within an organization. It is not only about implementing the necessary tools and protocols but also about fostering a work culture where cybersecurity is a fundamental aspect of the organization's core values. By doing so, we can ensure that every member of the organization is aware of the significance of cybersecurity and is equipped with the skills and knowledge to identify and mitigate potential threats.

We will cover the following topics:

- The importance of a cybersecurity-aware organizational culture

- Roles and responsibilities of different stakeholders

- Promoting shared responsibility for cybersecurity

We will examine why a culture that prioritizes cybersecurity is essential in today's business landscape. The significance of such a culture extends beyond the security or IT department; it is vital throughout all functions and levels of the organization. In a world where cyber threats have no borders or limitations to sectors, having an organizational approach to cybersecurity acts as a vital defense against potential breaches.

In the following section, we will discuss the various responsibilities and obligations of stakeholders in fostering a culture of cybersecurity. This includes individuals at all levels of the organization, from executives to front-line employees. Understanding these responsibilities is not only about holding people accountable but also about providing them with the necessary knowledge and tools to contribute to the organization's cybersecurity framework.

Finally, we will focus on strategies for promoting responsibility in the context of cybersecurity. This goes beyond training programs. It involves integrating cybersecurity awareness into everyday work processes and decision-making. The goal is to create an environment where cybersecurity is not viewed as a hindrance but, instead, as a component of organizational success and resilience.

This chapter aims to provide you with an understanding of the fundamentals of a cybersecurity-aware culture and suggest practical steps to establish and maintain it. The ultimate goal is to equip professionals with the knowledge required to transform their organization's approach to cybersecurity, moving it from being solely an IT concern to a shared commitment across the organization. The purpose of this chapter is to act as a guide to help create a culture where every individual comprehends, respects, and practices cybersecurity principles, which form the core defense against cyber threats.

The importance of a cybersecurity-aware organizational culture

Let's delve into why any modern organization must establish a culture that profoundly values cybersecurity. This section goes beyond perceiving cybersecurity as a requirement and emphasizes its importance as an integral component of a successful business strategy. In today's interconnected world, the consequences of cybersecurity breaches extend beyond the IT department and impact every aspect of a business.

We initiate our exploration by recognizing cybersecurity as an essential business priority, examining the rationale behind viewing cybersecurity not only as a matter itself but also as an essential function within the core operations of a business. This shift in perspective is vital for integrating cybersecurity into the strategy fabric, ultimately influencing decision-making at all levels. We will explore how cybersecurity influences customer trust, market reputation, and long-term financial well-being.

In moving on, it's critical to evaluate the risk and cost associated with cyber threats; our focus shifts to assessing what's at stake. This involves understanding the range of risks and costs associated with cyber threats, encompassing financial losses incurred from breaches to intangible yet equally significant expenses such as damage to brand reputation and loss of consumer confidence.

Organizations must conduct this assessment to fully understand how cyber incidents can impact their operations and growth. The role of leadership in shaping cybersecurity culture highlights the leaders' role in promoting a cybersecurity awareness culture. The commitment of leaders is essential for establishing a cybersecurity posture. This section will discuss how leaders can set the tone for cybersecurity by emphasizing its significance and integrating it into all business processes and decision-making.

By reading this section, the reader will understand why having a cybersecurity-aware organizational culture is not just an option but a necessity. The insights presented here will serve as the foundation for the following sections, which will delve into the responsibilities of stakeholders and strategies for fostering shared accountability in cybersecurity.

Understanding cybersecurity as a business imperative

Let us explore the indirect costs of cybersecurity breaches. Financially, companies face expenses for remediation and potential ransom payments and endure long-term consequences such as lost business opportunities and customer attrition. For instance, a small e-commerce company that suffers a data breach may experience revenue losses due to system downtime while also enduring damage to customer trust. These incidents underscore the financial vulnerability businesses confront in today's landscape. The Colonial Pipeline ransomware attack led to substantial indirect costs, including reputational damage and reduced consumer confidence, potentially affecting long-term customer relationships and market positioning. This incident highlights how cybersecurity breaches can extend far beyond immediate financial losses to encompass broader business impacts.

A lapse in security can affect customer perceptions, investor confidence, and overall brand image in industries such as finance or healthcare, where trust's intangible damages, such as tarnished reputation, can be even more detrimental than direct financial losses. For instance, a hypothetical breach in the security system of a healthcare provider could significantly damage the trust patients have in them, affecting their decision to use their healthcare services.

Another critical aspect is how cybersecurity plays a role in business operations. In today's interconnected business world, if there is a breach in one area, it can have effects throughout the organization. For example, if a manufacturing company's supply chain network is compromised, it can disrupt its operations and impact production schedules as customer commitments. This interconnectedness highlights the significance of having a reliable cybersecurity framework.

Companies that demonstrate superior cybersecurity measures can set themselves apart in industries where data security is of the highest concern. For instance, a financial services firm with a track record of safeguarding client data can leverage this achievement to attract and retain customers.

Regulatory compliance and its impact on business practices are another point. As governments worldwide tighten data protection and privacy laws, complying with these regulations has become not only an obligation but crucial for maintaining market credibility and avoiding penalties.

Lastly, we delve into the role of cybersecurity in fostering business growth and innovation. To successfully enter markets or create products, businesses must prioritize cybersecurity to safeguard their intellectual property and protect customer data securely.

All this highlights the importance of recognizing cybersecurity as a critical factor for any organization. It involves understanding how cybersecurity significantly impacts customer confidence, market reputation, adherence to regulations, competitive positioning, and fostering business innovation. This comprehension is essential for integrating cybersecurity into the core business strategy, ensuring it becomes an element of resilience and overall success.

Assessing the risks and costs of cyber threats

To begin the assessment, it is vital to identify the types of cyber threats that organizations may encounter. These threats can range from attacks such as phishing and malware to risks such as accidental data breaches or insider attacks. For example, a retail company might be vulnerable to point-of-sale malware, while a financial institution might have concerns about insider threats due to the sensitivity of its data.

Once the potential threats have been identified, the next step involves analyzing their impacts. This analysis considers costs, such as system repairs and data recovery, and indirect costs, such as legal fees, regulatory fines, and expenses related to loss of customer trust. For instance, if a law firm experiences a data breach, it will need to invest in securing its network after the breach. Still, it will also face the potential loss of clients due to concerns regarding attorney-client privilege.

Another essential aspect of assessing risks is understanding how cyber threats impact day-to-day operations. This involves considering the possibility of business disruptions and the associated downtime costs. For example, a manufacturing company could encounter halted production lines due to an attack, leading to operational losses.

The risk assessment should also consider the damage cyber incidents can cause to reputation. The erosion of customer trust and devaluation of brand value can have lasting effects that are challenging to measure but significantly affect the sustainability of a business. To illustrate this point, let's consider a consumer goods company that experiences a breach, resulting in decreased customer loyalty and negative brand perception. Furthermore, companies need to evaluate the implications of cyber threats on their industry and the overall global business landscape. For instance, if one central bank suffers from a cyber attack, it can lead to a loss of confidence in the banking sector and impact market stability.

Considering risks is also crucial. Failure to comply with data protection laws can result in fines and legal disputes. Businesses need to evaluate their cybersecurity measures concerning requirements to avoid facing legal consequences. Moreover, businesses need to consider the term's implications for cybersecurity. This encompasses the possibility of missed opportunities arising from a risk-averse culture born out of fear of cyber threats or the hindrance in adopting technologies due to security concerns.

Assessing the risks and costs associated with cyber threats is a process that demands a grasp of various aspects of cybersecurity. It involves more than quantifying losses; it also entails understanding how these threats can significantly impact a company's operations, reputation, compliance with regulations, and long-term strategic goals. By evaluating these risks, organizations can develop effective strategies to mitigate them and strengthen their overall cybersecurity stance.

The role of leadership in shaping cybersecurity culture

Leadership's responsibility starts with not only setting the tone from above but also leading by example by driving for an inclusive environment. For instance, when CEOs regularly discuss cybersecurity in meetings and company communications, it sends a message about its significance to the organization. This visible dedication can significantly impact how employees perceive and value cybersecurity, encouraging them to take it. Furthermore, leaders are accountable for incorporating cybersecurity into planning. This involves allocating resources towards cybersecurity initiatives and aligning them with business goals. For example, if the board invests in cybersecurity technologies, it demonstrates its commitment to safeguarding assets.

Training and awareness also fall within the domain of leadership. Leaders should advocate for and participate in cybersecurity training programs, as well as promote awareness among employees. It also sets a positive example for the rest of the company. For example, when executives actively participate in cybersecurity drills or training sessions, it demonstrates how important these activities are.

Leaders also play a role in creating an environment that promotes communication about cybersecurity concerns. They should establish channels where employees can report security issues without fear of retaliation. Encouraging employees to speak up about cybersecurity problems can lead to the detection of threats and foster a culture of vigilance.

Moreover, leaders should lead in developing policies and procedures that reinforce a culture focused on cybersecurity. This includes establishing guidelines for handling data, managing access controls, and following response protocols in case of a cybersecurity incident. For instance, implementing policies requiring password changes and two-factor authentication can significantly enhance the organization's security measures.

Leadership has a role in shaping a culture prioritizing cybersecurity. Through their actions, commitments, and policies, leaders influence how an organization approaches cybersecurity. By integrating cybersecurity into the organization's strategy, operations, and values, leaders ensure that it becomes everyone's responsibility and an integral part of the organization's foundation.

We will now shift our focus to the roles and responsibilities of different stakeholders. This critical part of the chapter acknowledges that cybersecurity is not solely the responsibility of IT professionals; rather, it is a shared duty that spans across roles within an organization.

Roles and responsibilities of different stakeholders

Each stakeholder, from the levels in the boardroom to every individual in the break room, plays a role in strengthening the organization's defenses against cyber threats.

We delve into the roles and responsibilities held by stakeholders within the cybersecurity framework. Our exploration extends beyond the individuals directly involved in IT security and encompasses how each employee, irrespective of their position, contributes to bolstering the organization's cybersecurity posture. Whether it's leaders setting goals and priorities for cybersecurity, the IT team implementing and managing security measures, or even front-line employees adhering to security policies, every role holds immense importance.

Moving forward, we learn about how to drive collaboration across departments and look into emphasizing the necessity for synergy among departments. Cybersecurity challenges are intricate and multifaceted, often demanding input and co-operation from areas within an organization, such as human resources, legal, finance, and operations.

In today's interconnected business environment, the security of our systems is only as strong as the link, which often extends to vendors, partners, and customers. Through this section, readers will gain an understanding of how a strong cybersecurity culture requires participation and collaboration from all internal and external stakeholders. It aims to demonstrate that effective cybersecurity is achieved through efforts across the organization, emphasizing that everyone has a role to play in protecting against cyber threats.

Defining stakeholder roles in cybersecurity

In the context of focusing on cybersecurity awareness, it is crucial to examine the individuals and groups involved within an organization. Let's take a look at the roles and responsibilities that each internal stakeholder holds in the overall cybersecurity narrative. By analyzing these roles, we shed light on the nature of engagement in cybersecurity and emphasize its crucial importance.

At the forefront of this exploration are the leaders who shape the organization's direction. They are more than figureheads; they are responsible for designing cybersecurity strategies that align with the overarching business objectives. For example, when a CEO actively supports cybersecurity initiatives, it signals a commitment that resonates throughout the organization, ensuring resources and attention.

Section 16 officers, under Sarbanes-Oxley, hold significant responsibilities to ensure transparent and ethical financial management, aligning closely with enhancing cybersecurity postures. They are mandated to report any stock transactions within the company promptly, reflecting a commitment to transparency and accountability. The SEC expects these officers to operate with utmost integrity, contributing to a corporate culture that values security and compliance. Their role is pivotal in establishing and maintaining trust, which is crucial for managing cybersecurity risks, as their decisions directly impact the organization's approach to safeguarding sensitive information and adhering to regulatory requirements.

In moving along this spectrum, IT professionals emerge as guardians of security. They oversee tasks such as implementing and managing cybersecurity measures while remaining vigilant against evolving threats. An IT manager might find themselves co-ordinating firewall deployments, conducting system audits, and leading responses to security breaches.

However, it is important to note that the stakeholders involved in cybersecurity efforts extend beyond those working in security or IT departments. Regardless of the job title, every individual has a part to play. By following cybersecurity practices, such as using passwords, being alert to phishing attempts, and promptly reporting suspicious activities, they can either strengthen or compromise the organization's defenses. Let's take a customer service representative who identifies and reports a phishing email; this person, too, plays a vital role in securing the organization. Always remember an organization's cybersecurity posture is only as good as its weakest link.

Beyond the IT sphere, leaders from departments also have a considerable impact. Heads of human resources, finance, and operations each bring their contributions to the table. HR leaders ensure that cybersecurity training and awareness are effectively spread throughout the workforce, and finance managers advocate for budget allocations towards cybersecurity initiatives.

The legal team, often working behind the scenes, becomes a link in the cybersecurity story. They make sure that the organizations cybersecurity policies comply with requirements and offer guidance during times of crisis. In case of any cybersecurity incident aftermaths, they skillfully navigate through the complexities and interact with regulatory bodies.

In this web of stakeholder involvement, the fusion of these roles into a unified cybersecurity strategy emerges as an essential theme. To achieve cybersecurity resilience, it is crucial to establish communication, implement shared training programs, and foster collaborative decision-making processes. Through this fusion of efforts, different departments can synchronize their actions and strengthen the organization's ability to safeguard against cyber threats.

Interdepartmental collaboration in cybersecurity

Now, let us explore the landscape of how various departments in an organization collaborate to strengthen cybersecurity measures. The cooperation between departments is crucial to taking an approach to cybersecurity when considering that threats often go beyond boundaries.

We begin by examining the role played by **human resources (HR)** in cybersecurity. HR departments have the responsibility of fostering a culture of awareness and understanding regarding cybersecurity among employees. This involves designing and implementing training programs that educate employees about cybersecurity practices. For example, HR managers may regularly conduct sessions to raise awareness among employees about the significance of passwords and how to recognize and report phishing attempts.

Moving on, we delve into the contributions made by the **finance department** to enhance cybersecurity. Finance professionals exert their influence by allocating resources in support of cybersecurity initiatives. Their role includes budgeting for security tools, personnel, and training. To illustrate this, a finance manager might advocate for an increased budget for cybersecurity to invest in solutions for detecting and responding to threats.

Legal departments also hold a position within the realm of cybersecurity. They ensure that the policies and practices related to cybersecurity comply with regulatory requirements. In the case of a cyber incident, legal teams guide organizations through compliance obligations as potential legal challenges. For example, they might navigate the complexities of data breach notification laws in the aftermath of a security breach.

Operational divisions such as **facilities and operations** also contribute to cybersecurity by implementing security measures. Elements such as access control systems, surveillance cameras, and visitor management protocols have an impact on cybersecurity. To illustrate this, the facilities team may enforce access controls to server rooms to prevent physical entry into critical infrastructure.

As we progressed further, we discovered the role played by **functional teams** during incident response. When a cybersecurity incident occurs, it is crucial for various departments to work together in a coordinated manner for a response. An instance of this is the creation of an incident response team that includes representatives from IT, legal, communications, and HR departments, working together to mitigate the impact of a breach and handle communication with any affected parties.

C-squite executives and **department heads** must champion cooperation and establish channels for communication and collaboration. Examples of this include meetings between departments to discuss cybersecurity strategies and conducting simulations involving multiple departments during incident response exercises.

Every department within the organization, including HR, finance, legal, and operations, plays a role in the cybersecurity framework. The central idea is that cybersecurity is an effort that thrives when different departments come together and leverage their strengths, working in harmony to safeguard the organization from changing threats.

Engaging external stakeholders in cybersecurity efforts

Let us go beyond the boundaries of an organization to examine how external stakeholders play a role in strengthening cybersecurity. These stakeholders, such as vendors, partners, and customers, are elements in the interconnected cybersecurity ecosystem. Our exploration begins by looking at relationships with vendors. Many organizations depend on vendors for services and technologies, ranging from cloud hosting to software development. These vendors may have access to data or systems, making their cybersecurity practices vital. For instance, consider a cloud service provider that hosts an organization's data; their security measures directly impact the security of that organization's data.

Partnerships are another aspect to consider. Organizations often collaborate with entities on projects or ventures. Such partnerships require a shared commitment to cybersecurity. For example, two financial institutions collaborating on a venture must ensure that their cybersecurity measures align to safeguard financial data.

The customer-stakeholder relationship is also highly significant. Customers trust organizations with their data. Expect it to be handled responsibly. Organizations must communicate their dedication to cybersecurity in order to build and maintain customer trust. In the healthcare industry, for example, patients have expectations when it comes to the security and confidentiality of their records. Any breach of trust can result in consequences. In 2021, the cost of data breaches in the healthcare sector reached an average of USD 9.3 million per incident, exceeding the general average cost of USD 9.23 million for data breaches across other industries. This highlights the particularly high financial impact of cybersecurity incidents within the healthcare field.

When it comes to compliance and regulations, external entities, such as industry watchdogs and government agencies, hold influence. Failure to comply with industry regulations or data protection laws can lead to penalties. For instance, the **General Data Protection Regulation** (**GDPR**) in the European Union imposes fines for mishandling customer data, which puts organizations under regulatory scrutiny.

Standards related to cybersecurity, such as ISO 27001, necessitate external audits. Organizations must demonstrate their adherence to these standards in order to earn trust. For instance, a cybersecurity company seeking ISO 27001 certification would undergo audits to validate its security measures and enhance its credibility in the market.

In today's digital era, organizations frequently interact with the public through media and online platforms. Managing perception and maintaining a reputation is of utmost importance. A single cybersecurity incident has the potential to significantly damage an organization's reputation. Organizations must proactively engage with the public in order to regain trust. For example, a social media company that experiences a data breach might use its platform to communicate about the incident and outline the measures taken to prevent any occurrences.

By collaborating and maintaining lines of communication with these stakeholders, organizations enhance their cybersecurity framework, establish trust, and showcase their dedication to protecting digital assets.

Next, we will delve into the topic of *promoting shared responsibility for cybersecurity*, where we will embark on a journey to cultivate a cybersecurity culture within our organization. We will learn about the importance of shared responsibility in strengthening an organization's cybersecurity posture, and we will explore the strategies and mechanisms that promote a sense of duty towards cybersecurity at all levels of the organization.

Promoting shared responsibility for cybersecurity

We will begin by delving into the strategies and initiatives that foster a culture where every individual within the organization, from executives to new employees, is fully aware of their role in ensuring cybersecurity. We will also draw inspiration from real-world examples of organizations that have successfully nurtured such cultures.

In moving forward on our journey, we come to the topic of how to build cross-functional cybersecurity teams, exploring the concept of forming functional teams dedicated to cybersecurity. These teams serve as strongholds for knowledge sharing, collaboration, and swift response to emerging threats. We will then explore how different industries have embraced these teams, the structures they have adopted, and the tangible impacts they have achieved.

Lastly, we will delve into how to measure and reinforce cybersecurity culture. You will learn how to highlight the role of metrics and continuous improvement in building a cybersecurity culture. We discuss how organizations can evaluate the effectiveness of their efforts to cultivate a culture of cybersecurity, identify areas for improvement, and implement strategies to maintain and strengthen awareness around cybersecurity. Additionally, we will provide methods for measuring progress and share success stories from organizations that have successfully developed sustained cybersecurity cultures.

Shared responsibility is essential for maintaining a cybersecurity posture. By fostering a culture where everyone takes ownership of cybersecurity regardless of their role, organizations can enhance their defenses and establish a lasting legacy of cyber resilience. The strategies and insights presented in this section can serve as a guide for organizations aiming to instill a culture that prioritizes vigilance in cybersecurity.

Creating a culture of cybersecurity awareness

Culture goes beyond being an aspiration; it is a practical necessity in the ever-changing landscape of cyber threats. Here, we will delve into the steps that need to be taken and the valuable insights into how to cultivate this type of culture.

The journey starts at the top, with leaders showing their commitment. Executives and managers should lead by example, demonstrating a dedication to cybersecurity. They can actively participate in cybersecurity training sessions alongside their teams, reinforcing the importance of security being everyone's responsibility. This commitment should also be visible through resource allocation and strategic decisions, highlighting the organization's dedication to cybersecurity.

Education forms the foundation for cybersecurity awareness. However, it's not sufficient to provide knowledge. Consider incorporating exercises that mimic real-life scenarios into your training programs. For instance, simulate phishing attacks to help employees recognize and respond effectively to attempts. These hands-on exercises equip employees with skills that make them more resilient against threats.

Cyber threats evolve rapidly, necessitating learning as a practice. To maintain vigilance, establish a culture where continuous learning is encouraged and valued. To promote cybersecurity awareness within the organization, it is important to conduct sessions where employees can learn about emerging threats and best practices. Online resources and newsletters can also be shared to keep everyone updated. For example, quarterly briefings can be organized as forums for employee engagement and information sharing.

To ensure that cybersecurity awareness is taken seriously, accountability measures should be put in place. Employees should be required to complete cybersecurity training modules, and their understanding of the subject should be assessed. Non-compliance should have consequences, such as restricting access to systems until the training requirements are met. This encourages compliance and also highlights the importance of cybersecurity within the organization.

Effective communication channels are crucial for reporting security concerns. Clear and accessible channels should be created so employees can report activities or security incidents. Whistleblower hotlines, email addresses, and incident reporting forms are tools that should be regularly communicated and encouraged.

Cybersecurity training materials should cater to all departments and job roles within the organization. Each department can receive training tailored to their functions. For instance, HR teams can receive training on data privacy regulations with an emphasis on safeguarding employee data. Similarly, marketing departments can learn about protecting customer data and recognizing social engineering attempts.

Lastly, integrating cybersecurity into the culture is essential for long-term success. It involves creating a mindset where every employee understands their role in maintaining cybersecurity measures throughout operations.

The main objective is to incorporate cybersecurity awareness into the culture of the organization. This can be achieved through reminders for recognizing and rewarding security practices and maintaining open lines of communication. The goal is to create an environment where cybersecurity is seen as a part of operations rather than a hindrance. By implementing these steps, you can establish a culture of cybersecurity awareness that permeates throughout your organization, bolstering its ability to withstand cyber threats.

By leveraging these insights and real-life examples, you can empower your organization to navigate the intricate world of cyber threats. It's crucial for every member of the team to actively defend your assets, fostering a sense of resilience and confidence within your organization.

Building cross-functional cybersecurity teams

Let's embark on a journey to delve into the world of functional cybersecurity teams. These teams play a role in combating the changing landscape of cyber threats and also foster collaboration and knowledge sharing across different domains within your organization.

Cross-functional teams bring together experts from different backgrounds, such as IT, compliance, legal, and risk management, to collectively address cybersecurity challenges. The aim is to combine knowledge and skills to create a force capable of tackling complex threats.

Establishing a structured, functional cybersecurity team is crucial. It involves defining roles and responsibilities that align with the required expertise. For instance, you could designate a threat intelligence analyst to monitor emerging threats or assign a legal compliance specialist to ensure adherence to data protection regulations.

Sharing knowledge forms the backbone of functional teams. Encourage team members to exchange insights, best practices, and up-to-date threat intelligence. This can be facilitated through team meetings, collaborative platforms, and repositories for storing knowledge. For example, you can create a platform where team members can share relevant articles and alerts about threats and lessons they have learned.

Conduct training exercises that simulate real-world cyber incidents to prepare your team. These exercises should involve simulating cyberattacks, data breaches, or phishing attempts. Include team members from different disciplines in the response. By doing so, you help team members understand their roles better and develop a co-ordinated strategy for responding.

Furthermore, developing playbooks that outline the actions each team member should take during a cyber incident. Tailor these playbooks to address the risks and challenges faced by your organization. For instance, create a playbook that guides the response to a data breach with input from legal, IT, and PR teams.

Moreover, effective communication is crucial for cross-functional teams to succeed. Establish communication channels and protocols for sharing information during an incident. Clearly define roles, responsibilities, and reporting lines to ensure seamless information flow. Consider utilizing collaboration tools that facilitate real-time communication and document sharing.

Establish **key performance indicators** (**KPIs**) aligned with your cybersecurity goals to assess how effectively your cross-functional cybersecurity team is performing. Keep track of metrics such as how incidents are responded to, the rate at which incidents are resolved, and the level of participation in knowledge sharing. Make it a habit to regularly review these metrics and make any adjustments to improve your team's performance.

By using these insights and examples, you can empower functional cybersecurity teams that can handle cyber threats and promote a culture of collaboration and knowledge sharing within your organization. These teams act as a line of defense, ensuring that cybersecurity efforts go beyond departmental boundaries and become a collective effort.

Measuring and reinforcing cybersecurity culture

While it's crucial to establish a culture of cybersecurity awareness, it's equally vital to have ways to measure its effectiveness and continuously enhance it. Here, we explore implementing a data-driven approach to achieve these goals.

To begin this journey, it's essential to establish metrics that align with your organization's objectives for cybersecurity culture. These metrics can cover aspects such as the completion rate of cybersecurity training or the number of reported security incidents. For example, you can measure the percentage increase in employees completing security training modules year after year.

Surveys are tools for understanding employee perspectives on cybersecurity within your organization. Design surveys that ask about awareness levels, attitudes, and behaviors related to cybersecurity. Use a rating scale to quantify the responses. For instance, you can ask employees to rate their confidence in identifying phishing emails on a scale from 1 to 5.

Analyzing incident reports can provide insights into the effectiveness of your organization's cybersecurity culture. Look for trends in incident reporting, such as types of incidents reported and reporting speed. For instance, if employees start reporting activities quickly, it can indicate that the cybersecurity culture has improved.

Consider implementing a rewards system to acknowledge and motivate employees who consistently report security concerns in training modules on time or excel in phishing exercises. This could include offering recognition or small incentives.

Utilize the data gathered from metrics, surveys, and incident reports to identify areas where your cybersecurity culture can be improved. Regularly review and update cybersecurity training materials, awareness campaigns, and incident response procedures based on gained insights. For example, if survey results reveal that a specific training module is confusing for employees, make revisions to make it more user-friendly.

Share the results of cybersecurity culture assessments with all employees to foster trust and promote shared responsibility. Highlight success and improvements during company meetings to demonstrate the organization's commitment to cybersecurity. For instance, emphasize how employee awareness has contributed to a decrease in phishing attempts.

Measuring and reinforcing cybersecurity culture requires effort. Regularly assess the impact of your initiatives. Adapt strategies accordingly. Always listen to your employees' feedback to ensure that your initiatives align with their needs and truly resonate with them.

By putting these strategies into action and utilizing insights derived from data analysis, you can effectively strengthen your organization's cybersecurity culture. This approach, driven by data, not only helps quantify the progress achieved but also guarantees the resilience of your cybersecurity culture against evolving threats.

Summary

This chapter emphasized the significance of integrating cybersecurity into the organizational culture, stressing its importance at all levels and not just within IT or security departments. It proposed an approach whereby cybersecurity is tightly integrated into an organization's core values, enabling everyone to address cyber threats and highlighting the need for team effort in defending against cyber threats; this promotes responsibility among stakeholders to go beyond boundaries.

Furthermore, the chapter delved into the importance of considering cybersecurity as a business priority by discussing its impact on customer trust and overall financial well-being. It showcased how leadership plays a role in nurturing a culture centered around cybersecurity through integration and decision-making. The content moved on to outline stakeholder responsibilities, advocating for collaboration across departments and extending cybersecurity commitment to parties such as vendors and customers. By offering strategies to foster, assess, and improve cybersecurity culture, we saw that maintaining an alert and collectively dedicated environment is vital for cybersecurity defenses.

12
Collaborating with Industry Partners and Sharing Threat Intelligence

Collaboration and sharing threat intelligence are vital in establishing a successful cybersecurity defense. This chapter delves into these concepts, explaining how they form the foundation of strategies for defending against cyber threats and continuous security posture improvements. With the rapidly evolving threat landscape, the importance of an effective cybersecurity defense becomes more evident. This chapter examines the mechanisms and advantages of intelligence sharing, trust building, and collective responsibility in cybersecurity.

We'll cover the following topics:

- The importance of collaboration and threat intelligence sharing
- Building trust and maintaining confidentiality in information sharing
- Leveraging shared threat intelligence for improved security
- Promoting shared responsibility for cybersecurity

The initial section underscores the criticality of exchanging information on cyber threats, vulnerabilities, and exploits among organizations. We'll explore examples and theoretical frameworks that highlight the benefits of being watchful in identifying and countering cyber threats. This part aims to illustrate the advantages of adopting an approach to threat intelligence.

Subsequently, we'll delve into addressing the intricacies of building trust within cybersecurity. This section underscores how trust and confidentiality are vital for successfully sharing intelligence. By examining models and protocols crafted to protect data, we can lay a roadmap for fostering trust among partners within the cyber threat intelligence landscape.

The following section delves into implementations of shared intelligence. This section demonstrates how companies can incorporate and put shared data into practice to strengthen their security stance. From searching for threats to reacting to incidents, the emphasis is on turning intelligence into strategies that prevent and counter cyber threats.

Moreover, the final section promotes the belief in shared responsibility within cybersecurity. Effective defense against cyber threats is not just up to IT departments but requires a culture of awareness and responsibility across the organization. By showcasing projects and partnerships that embody this shared responsibility, this section aims to encourage an approach to cybersecurity.

Throughout this chapter, our goal is to provide you with the knowledge and resources needed to participate in and actively contribute toward a secure digital space through cooperation and shared information.

The importance of collaboration and threat intelligence sharing

The key to defending against cyber threats is exchanging knowledge, insights, and data across different organizations. This section highlights why working together and sharing threat information are not just choices but essential elements in today's cybersecurity landscape.

We'll explore the advantages of sharing threat intelligence, such as improving your organization's **mean time to detect (MTTD)** and **mean time to respond (MTTR)** metrics. MTTD is the average time it takes for an organization to detect a cyber threat after it has occurred, while MTTR measures the average time required to respond to the detected cyber threat. Using examples and theoretical perspectives, we aim to show how collaboration can turn knowledge into a shield against cyber attackers. This opening sets the scene for understanding how shared intelligence serves as a tool to help cybersecurity experts predict, identify, and address threats more efficiently than ever before. Through this conversation, we'll establish the foundation for discussions on fostering trust, utilizing intelligence effectively, and promoting responsibility – each playing a crucial role in enhancing cybersecurity collaboration.

The imperative for collaborative defense

Collaboration among organizations and industries has become essential to the battle against cybercrime. It fosters a front that taps into expertise and resources. For example, the joint efforts to combat the DDoS attacks by the Mirai botnet demonstrated the strength of working for defense. **Internet service providers (ISPs)**, cybersecurity companies, and impacted organizations shared up-to-date information on attack methods and ways to mitigate them, showcasing how collaboration can reduce the impact of threats.

Another case in point is the partnership between cybersecurity firms and government agencies, which has been crucial in uncovering and dismantling cybercriminal networks. The successful takedown of the Emotet botnet in 2021 is an instance where global law enforcement agencies and private partners collaborated to disrupt what was deemed one of the most dangerous malware strains globally. This operation emphasized the significance of cross-industry cooperation in addressing worldwide cyber risks.

Groups such as the **Cybersecurity and Infrastructure Security Agency** (**CISA**) in the US and the **European Union Agency for Cybersecurity** (**ENISA**) in Europe help companies and government agencies share information about cyber threats. To use these groups effectively, companies and agencies can join their networks or partnerships. This lets them get alerts about new threats and advice on how to protect themselves. Sharing their own experiences with threats can also help others prepare and respond better.

Furthermore, alliances specific to industries such as the **Automotive Information Sharing and Analysis Center** (**Auto ISAC**) illustrate sector-focused collaboration. By exchanging intelligence on threats related to technology and cybersecurity, member companies can strengthen their defenses against targeted assaults. These instances underscore how collaborative defense strategies are essential in today's interconnected world of cybersecurity.

Mechanisms of threat intelligence sharing

- **Established systems for information exchange**: Successful threat intelligence sharing is built on the foundation of established systems and platforms, such as the **Financial Services Information Sharing and Analysis Center** (**FS ISAC**). This setup allows organizations within the same sector to exchange insights on threats and vulnerabilities, enhancing security measures through collective knowledge.

- **Industry-specific sharing**: By focusing on industry-specific intelligence, like that of FS ISAC, the shared information is both relevant and actionable for all members, ensuring that the intelligence has practical applications.

- **Global collaboration**: The **Global Cyber Alliance** (**GCA**) exemplifies large-scale cooperation by uniting law enforcement agencies and research institutions. This alliance offers tools, resources, and intelligence to systematically address cyber risks and improve internet security, with initiatives such as a secure DNS service protecting users from malicious websites.

- **Threat intelligence platforms** (**TIPs**): Platforms such as the **Malware Information Sharing Platform** (**MISP**) and Threat Sharing facilitate the exchange, storage, and correlation of intelligence on cyber threats. These platforms allow organizations to share **indicators of compromise** (**IoCs**) and **tactics, techniques, and procedures** (**TTPs**) used by adversaries, enhancing collective defense capabilities.

- **Implementation and customization**: Utilizing TIPs such as MISP involves installing the platform on a server and customizing it to fit the organization's sharing preferences, user roles, and security protocols. This enables effective handling, exchange, and response to cyber threat intelligence.

- **Community engagement and event creation**: Organizations can contribute to the cybersecurity community by creating "events" in platforms such as MISP to outline specific threats or attack patterns. This information, ranging from phishing campaigns to malware signatures, can be shared with select partners or groups, fostering a collaborative approach to threat detection and response.

- **Benefiting from shared intelligence**: Subscribing to events shared by others allows organizations to access vital intelligence that enhances their defense mechanisms, such as updating security measures and conducting targeted scans for known threats.

- **Enhancing collective understanding**: MISP's capability to automatically link indicators across events aids organizations in recognizing attack patterns and connecting incidents. This collective insight into the threat landscape not only strengthens security measures but also facilitates uncovering new attack methods or command and control servers, benefiting the wider community.

Best practices in collaboration and sharing

To effectively share threat intelligence, it's essential to follow best practices. I have listed a few here:

- **Building trust**: This is critical in collaboration and involves creating a setup where everyone trusts the honesty of the information shared and the confidentiality maintained. Agreements such as **memorandum of understanding (MOUs)** can help set expectations and responsibilities for each party involved, ensuring that intelligence is handled securely and appropriately.

- **Reciprocity**: This also plays a role in maintaining networks. Sharing information should be a two-way process that requires participation and contributions from all members. For example, if an organization discovers a phishing technique targeting their industry, sharing information about the attack helps others defend against it and encourages mutual sharing, enhancing the overall pool of intelligence.

- **Integrating shared intelligence into security operations**: This shifts from reception to use. When an organization receives intel on a malware variant, promptly updating antivirus signatures and informing the SOC team enables defensive actions. This practical application of shared intelligence strengthens an organization's cybersecurity stance, showcasing the advantages of collaboration and threat intelligence sharing.

The examples outlined here for effective threat intelligence sharing, including building trust, ensuring reciprocity, and integrating intelligence into security operations, are crucial but not comprehensive. The field of cyber threat intelligence sharing is dynamic, emphasizing the necessity for secure information exchange and mutual contributions. Trust and reciprocity lay the groundwork for this exchange, while the practical application of shared intelligence showcases the tangible benefits of collaboration. These principles significantly strengthen an organization's cybersecurity capabilities, underlining the collective enhancement of detection and response to cyber threats.

Collaboration and sharing threat intelligence play roles in today's cybersecurity landscape. They bring advantages, such as pooling knowledge and resources to bolster an organization's ability to effectively identify, prevent, and address cyber threats. This section has explored platforms and frameworks for sharing information, highlighting collaborative initiatives. By grasping the dynamics of collaboration and the processes involved in sharing threat intelligence, organizations can significantly enhance their cybersecurity defenses, fortifying them against evolving threats.

The skills conveyed in this section are essential for cybersecurity professionals aiming to harness their capabilities thoroughly. You have gained insights into the significance of forging partnerships across organizations and sectors, enabling a more unified response to cyber threats. Furthermore, proficiency in navigating and utilizing threat intelligence platforms equips professionals with the tools for collaboration. These competencies boost organizations' security measures and strengthen the broader cybersecurity community's resilience. By embracing collaboration principles and shared intelligence practices, cybersecurity experts are better equipped to tackle the complexities of an evolving threat landscape, ensuring a better digital future for everyone.

Now that we have understood the importance of collaboration and threat intelligence, we must learn about establishing trust among partners. That is what we are going to cover in the following section.

Building trust and maintaining confidentiality in information sharing

Establishing trust among partners is a process for securely exchanging sensitive information. This section explores building trust, from screening partners to maintaining trust through ethical sharing practices. It also underscores the significance of upholding confidentiality to safeguard the interests of all parties involved and maintain the integrity of shared information.

The fine line between transparency and confidentiality will also be examined, emphasizing strategies for sharing intelligence without compromising security or exposing sensitive data. Frameworks such as the **Traffic Light Protocol** (**TLP**) play a role by providing guidelines for classification and sharing. Furthermore, implementing encryption and secure communication channels will be discussed as it can safeguard shared intelligence.

Legal and compliance considerations are vital as they establish a framework for information sharing. Involving teams in drafting information-sharing agreements ensures compliance with requirements and strengthens the foundation of trust.

This section highlights the roles of trust and confidentiality in nurturing an efficient cybersecurity ecosystem. By navigating these principles, organizations can contribute and benefit from collective defense against cyber threats, bolstering the overall security stance of the digital community.

Establishing trust among partners

Establishing and nurturing trust requires effort from all parties involved. Initially, fostering trust entails assessing collaborators to ensure their dependability and the alignment of their cybersecurity protocols with their own. This evaluation typically involves gauging the maturity and reputation of their cybersecurity procedures within the field. Guidelines such as those provided by **Information Sharing and Analysis Organizations (ISAOs)** offer direction on cultivating trust among entities and facilitating information-sharing relationships based on respect and shared goals.

Organizations frequently enter into agreements to solidify trust that delineates engagement terms, responsibilities, and expectations for each party. Such agreements, such as **non-disclosure agreements (NDAs)**, represent a formal commitment to safeguarding the confidentiality and integrity of shared information. They serve as a cornerstone for trust, assuring all parties that shared insights will be handled carefully and kept confidential.

Consistent communication and transparency regarding how shared information is utilized, as well as feedback mechanisms, can help sustain and potentially enhance trust over time.

Sharing experiences where partners' information has helped prevent a cyber threat can confirm the trust placed in the network.

Maintaining confidentiality in information sharing

Ensuring confidentiality when sharing threat intelligence is crucial for protecting the interests of all parties involved and maintaining their willingness to exchange information. It's crucial to enforce controls on access and establish protocols for handling sensitive data. We will discuss a few ways we can do this:

- **TLP**: This is widely recognized for promoting confidentiality by categorizing information based on its sensitivity level and providing guidelines on who can access it. The TLP system uses colors to indicate the sensitivity of data:

  - **TLP;RED**: This is for sensitive information that's restricted to specific individuals or groups within an organization

  - **TLP;AMBER**: This is for moderately sensitive data that's shared with a limited audience who can act upon it

  - **TLP;GREEN**: This is for less sensitive information that's shared within the organization or with external partners under certain restrictions

  - **TLP;WHITE** (or **TLP;CLEAR**): This is for information that can be publicly shared without limitations

When using TLP, it is important to assign a color code to information when it is first shared based on its sensitivity and the impact it will have if it is disclosed without authorization. For instance, if an organization discovers a cyber threat and puts together an intelligence report, it will mark the report with a TLP color indicating who can see it. If the report includes vulnerability details that could cause harm if leaked, it might be labeled as TLP;RED or TLP;AMBER, thus restricting access to those directly involved in handling the threat. This method not only safeguards data from being widely exposed but also ensures that crucial intelligence is shared with the right people who can take action, strengthening the security of all parties involved.

- **Utilizing encryption and secure communication channels**: This plays a role in upholding confidentiality. Whether you're transmitting information via platforms or informal means, data must be encrypted during transit and storage to prevent entry. Moreover, employing tools and technologies that facilitate authentication and verification adds a layer of protection against potential compromises of sensitive information.

- **Regularly reviewing data and access permissions**: This is crucial for adapting to evolving threat landscapes and organizational changes. This ongoing evaluation helps reduce the risk of data breaches or leaks. For instance, promptly reassessing employees' access to shared threat intelligence when they leave the organization or switch roles can prevent disclosures.

Balancing transparency and confidentiality

Maintaining a balance between openness and confidentiality is key for collaboration in sharing threat intelligence. It's essential to be transparent about where the information comes from, the methods used for analysis, and the reasoning behind threat assessments to enhance trust and value in shared data. However, handling transparency is essential to prevent vulnerabilities or sensitive operational details from being exposed. Here are a few strategies to employ:

- **Effective communication**: This plays a role in setting boundaries that uphold confidentiality while encouraging information exchanges.

- **Anonymizing data**: Anonymized data allows organizations to share insights gleaned from datasets without revealing the identities of individuals within the data.

- **Collectively sharing intelligence**: This can enable organizations to contribute to security efforts without disclosing their weaknesses.

- **Information-sharing agreements**: Engaging compliance teams in crafting information-sharing agreements can ensure that these practices align with regulations and company policies. This does not safeguard the organization legally. It also strengthens partner trust by showcasing a dedication to ethical and responsible sharing protocols.

As we conclude this section, where we discussed the importance of trust and confidentiality in information sharing, it's clear that these aspects are crucial to collaboration in cybersecurity. This examination has underscored the role of building trust among partners and implementing robust measures to safeguard the confidentiality of shared threat intelligence. By delving into frameworks such as TLP and utilizing NDAs, we've seen how structured approaches can create a secure environment for sharing sensitive information. Emphasizing encryption and secure communication channels further underscores maintaining confidentiality and ensuring that shared data remains protected from access.

The knowledge you've gained from this section is crucial for any cybersecurity professional seeking to participate in collaborative threat intelligence sharing. You have been equipped with strategies for establishing a trustworthy foundation that fosters the open exchange of vital cybersecurity information. You have also acquired insights into crafting and enforcing agreements for safeguarding shared intelligence, striking a balance between transparency and security. These skills are essential in nurturing a cybersecurity environment where trust and confidentiality lay the groundwork for resilience against cyber threats.

Using shared threat intelligence effectively is essential for organizations seeking to enhance their cybersecurity defenses. The following section delves into the benefits of leveraging shared intelligence in transforming security practices, refining decision-making processes, and expediting incident response efforts. Organizations can boost their awareness of threats by integrating this knowledge into security operations.

Leveraging shared threat intelligence for improved security

The practical applications of shared intelligence are vast, encompassing activities such as bolstering alert systems with real-time threat updates and prioritizing patch management based on vulnerability information. Lastly, the collaborative aspect of shared intelligence extends to incident response scenarios, where insights from breaches and recovery endeavors can guide organizations during crises. This collective expertise not only aids in mitigating threats but also fosters a culture of heightened security consciousness within organizations.

Through discussions and examples, this section aims to demonstrate how strategically leveraging shared threat intelligence can result in security enhancements. It underscores the significance of an approach in translating shared knowledge into security improvements, thereby fortifying organizational cyber resilience amid an ever-changing threat landscape.

Integrating shared intelligence into security operations

Utilizing shared threat intelligence in security operations significantly boosts an organization's capability to tackle cyber threats proactively. The first step involves promptly setting up mechanisms for receiving, analyzing, and applying intelligence. The **Security Information and Event Management (SIEM)** system is a hub for handling the diverse information that makes up threat intelligence. To effectively leverage threat intelligence within a SIEM, the first step involves configuring the SIEM to receive threat feeds from external sources. These feeds include data such as IoCs and TTPs used by attackers, and other relevant threat indicators. The integration process typically involves establishing API connections or importing data in standardized, machine-readable formats such as STIX/TAXII, ensuring that the SIEM system can efficiently retrieve and process the latest cyber threat intelligence. Utilizing STIX/TAXII not only streamlines the process of sharing and receiving threat intelligence across different platforms but also enhances an organization's ability to quickly adapt its defenses against evolving cyber threats, ensuring more robust and timely cybersecurity responses. Once the SIEM has been configured to receive these feeds, it can correlate this intelligence with internal event data collected from sources within the organization's network, such as logs from firewalls, intrusion detection systems, and endpoint protection platforms. This correlation allows for efficiently identifying potential threats by comparing current network activities with known malicious behaviors and indicators.

For example, suppose the SIEM system detects network activity directed toward an IP address known for hosting a command and control server associated with a type of malware. In that case, it can trigger an alert for investigation. Additionally, SIEM platforms empower organizations to rank incidents according to their seriousness and relevance regarding threat intelligence. By utilizing threat intelligence to assess the impact of an alert, security teams can concentrate on addressing the critical issues first. This may involve configuring the SIEM to assign priority levels to alerts linked to IoCs from breaches or threats specifically pertinent to the organization's industry or technology environment.

To delve deeper into this topic, let's explore how shared intelligence plays a role in combating phishing attacks. By integrating intelligence on phishing tactics and IoC systems into filtering mechanisms, organizations can proactively block emails, significantly reducing the likelihood of attacks. This operational application of intelligence not only safeguards individual organizations but also strengthens community defense by providing insights into the effectiveness of mitigation strategies.

Furthermore, the importance of threat intelligence in improving vulnerability management cannot be overstated. Organizations can prioritize patching efforts based on trends and threat actor behaviors by leveraging insights from shared intelligence sources. This targeted strategy for managing vulnerabilities ensures that resources are used efficiently by addressing risks posed to the organization.

The importance of sharing threat information goes beyond exchanging IoCs. It also involves sharing insights on how attacks are carried out, strategies to mitigate them, and ways to recover from them. This is what we will be learning about in the following section.

Collaborative incident response and recovery

Working together collaboratively during a security breach can help minimize the downtime and financial losses caused by the attack. For example, in the case of an incident, receiving intelligence from organizations that have effectively dealt with situations can offer valuable advice on containing and recovering from such attacks.

Online platforms and communities play a role in facilitating knowledge exchange. Platforms such as the **Cyber Threat Alliance (CTA)** enable members to share analyses of malware and cyberattack campaigns, promoting a collaborative approach to dealing with security incidents. These platforms provide organizations with access to various experiences and expertise, empowering them to respond confidently and effectively. The CTA facilitates information sharing among its members regarding malware analyses, cyberattack campaigns, vulnerabilities, and threat indicators. This sharing process is structured within a framework to ensure that the information that's exchanged is both timely and actionable. Members actively contribute by sharing their discoveries in a format that then becomes accessible to others on the platform. The collaborative nature of this model aims to prevent cyber attacks and minimize their impact globally by ensuring that all members have up-to-date threat intelligence at their disposal. The CTA operates based on principles of reciprocity and a mutual dedication to enhancing ecosystem security. Members are not encouraged to share IoCs but provide contextual information about attacks, including TTPs employed by attackers to enhance collective knowledge of cyber threats. This inclusive approach enables members to defend against known threats while also preparing for emerging tactics. Another example of a collaborative effort in incident response and recovery is the **Global Cyber Alliance (GCA)**. The GCA brings together international partners from various sectors, including government, law enforcement, and the private sector, to work on initiatives aimed at reducing cyber risks and improving internet security worldwide. Unlike the CTA, which focuses on sharing threat intelligence among its members, the GCA aims to implement practical solutions that prevent cyber threats. For instance, the GCA has developed tools such as the Quad9 DNS service, which blocks known malicious domains, thereby preventing attacks before they happen. Additionally, the GCA's project on securing email systems against phishing attacks is another example of how collaborative efforts can lead to the development of concrete tools and protocols that enhance cybersecurity. The GCA's approach emphasizes the power of collective action in developing and deploying solutions that have a direct impact on reducing the frequency and severity of cyber incidents, thereby aiding in quicker recovery and minimizing losses.

Working together to gather and analyze cybersecurity information can significantly boost an organization's ability to fend off cyber attacks. By incorporating shared knowledge into security practices, organizations can enhance their understanding of threats, improve how they detect and respond to them, and effectively prioritize areas of vulnerability. Leveraging a pool of information empowers organizations to be more proactive in defending against cyber attacks, reducing the chances of breaches and lessening the impact when breaches occur. Examples illustrate how collaborative efforts can help identify, address, and recover from cyber incidents, demonstrating the advantages of using shared intelligence to bolster cybersecurity defenses.

The expertise gained through this exploration is invaluable for cybersecurity professionals looking to enhance their organization's security readiness. You've acquired insights into integrating shared threat intelligence into your security frameworks, from setting up SIEM systems to leverage real-time threat data to utilizing shared insights for vulnerability management and coordinated incident response efforts.

Furthermore, promoting a culture of security awareness within an organization plays a role in making the most of shared intelligence resources by underscoring the importance of every staff member in safeguarding against cyber threats. These skills equip organizations to address cybersecurity issues more efficiently and lay a solid foundation for adapting to future threats in an ever-evolving digital landscape, ensuring resilience against upcoming cybersecurity challenges.

Now that we have seen how cybersecurity professionals can enhance their organization's threat readiness, we must understand that cybersecurity isn't just up to IT departments or security experts alone. It's an effort that involves everyone in an organization and the wider digital world. The following section delves into the idea of shared responsibility in cybersecurity.

Promoting shared responsibility for cybersecurity

This section emphasizes how working together can significantly boost protection against cyber dangers. A stronger and more secure cybersecurity stance can be established by involving individuals, organizations, and governments. It's not about collaborating; it's about fostering a culture where each party knows their part in protecting assets. Real-life examples will showcase how shared responsibility can be implemented within organizations and alongside partners. By promoting a culture of security awareness and proactive measures, we can build a space for all. This approach does not reduce risks. It also ensures quick and coordinated responses to new threats, showcasing the strength of unity in cybersecurity.

Cultivating a culture of cybersecurity awareness

Collaborating on threat intelligence plays a role in fostering a security-focused culture within a company. It all begins with sharing details about threats with the security teams, who then analyze the information to pinpoint risks. Subsequently, these teams translate threat data into insights that can be easily grasped by all employees. For instance, in the event of uncovering a phishing scam, security teams can develop resources outlining the scheme, its warning signs, and the necessary steps for employees to take if they encounter such an email. This proactive approach to education ensures that everyone remains well-informed and alert, contributing to the organization's defense strategy.

To implement this effectively, conducting training sessions and workshops is vital for educating staff on threats and promoting online practices. These sessions might involve simulated phishing exercises where employees are presented with phishing emails to assess and improve their ability to identify risks. Feedback gathered from these drills can guide training initiatives for relevance and effectiveness. Additionally, establishing a hub for user security advisories and guidelines based on up-to-date threat intelligence can serve as a resource for employees. This repository should be easily accessible and regularly updated to encourage awareness among staff regarding threats. Furthermore, leveraging communication channels, such as intranets, emails, or messaging apps, to disseminate alerts regarding emerging threats can significantly enhance awareness levels across the organization.

For example, suppose a malware attack spreads using engineering methods. In that case, a brief notification can be sent to all team members with information about the attack, how it works, and tips to prevent them from being targeted. This immediate communication highlights the significance of being aware of security and guarantees that employees are equipped with the information to protect themselves and the company. By integrating collective threat intelligence into employee training and communication initiatives, companies can enable their staff to be knowledgeable and alert and play a role in their cybersecurity protection plan.

The core of responsibility in cybersecurity is rooted in establishing a culture of awareness across the organization. This process commences with training initiatives that educate all staff on cybersecurity practices, such as identifying phishing scams, emphasizing the significance of robust passwords, and ensuring secure data handling. A practical method to enhance this awareness is phishing drills, which enable employees to recognize and appropriately respond to emails.

Organizations should cultivate an environment where cybersecurity remains a priority. Providing updates on emerging threats and reinforcing security measures serve to keep cybersecurity at the forefront of employees' minds. One practical approach involves utilizing newsletters to showcase cyber incidents relevant to the organization and offering guidance on addressing similar threats.

Leadership plays a role in championing this mindset. When executives exhibit a dedication to cybersecurity, it sets a precedent for the organization. For instance, CEOs engaging in cybersecurity training alongside their team members reinforces the notion that safeguarding assets is a collective responsibility.

Engaging in public-private partnerships (PPPs)

PPPs are crucial for fostering a sense of shared responsibility in cybersecurity. These collaborations harness the strengths and resources of both private sectors to bolster cybersecurity resilience on both national and global scales. For instance, in the United States, the **National Cybersecurity and Communications Integration Center** (**NCCIC**) works hand in hand with sector partners to exchange threat information and coordinate responses to security incidents.

By working with these partnerships, we can establish standards and frameworks that benefit a diverse array of stakeholders. One tangible illustration is the collaboration between government bodies and private enterprises to develop the **NIST Cybersecurity Framework**, which offers guidance for organizations seeking to effectively manage and mitigate cybersecurity risks.

Furthermore, PPPs can facilitate joint simulations that assess the preparedness of private entities for cyber threats. Exercises such as Cyber Storm, organized biennially by the Department of Homeland Security, prove invaluable in pinpointing any deficiencies in response strategies and enhancing coordination among all participants.

Leveraging technology for collective defense

The role of technology is crucial in encouraging a sense of shared responsibility toward cybersecurity. Organizations can collectively strengthen their defense capabilities by utilizing platforms and tools. We will discuss a few strategies here:

- **Use of cloud services**: Cloud services that offer advanced security features benefit **small and medium-sized enterprises** (**SMEs**) that may not have the resources to implement these measures on their own.

- **Adoption of standards for security technologies**: This promotes interoperability and the exchange of threat intelligence. One such standard is the **Security Assertion Markup Language** (**SAML**), which enables sharing authentication and authorization data between entities, enhancing the security of online transactions across various domains.

- **Open source and shared responsibilities**: Technology also promotes shared responsibility through projects such as open source security tools. Initiatives such as **OWASP Zed Attack Proxy** (**ZAP**) provide organizations with effective tools to detect vulnerabilities in their web applications, showcasing how collective efforts can result in reliable security solutions for everyone.

This section has underscored the importance of fostering a culture of cybersecurity awareness within companies engaging in partnerships between the private sectors and using technology as a united defense strategy. Each approach emphasizes that cybersecurity is not about IT departments but is a duty involving every individual organization and government body.

By delving into instances such as simulated phishing drills and collaborative efforts in creating the NIST Cybersecurity Framework, we have witnessed how these methods can be implemented to create a digital landscape. The knowledge gained from this section is crucial for cybersecurity experts and all members of an organization. You have acquired insights into developing training programs that enhance employees' awareness and readiness, thereby strengthening the organization's ability to combat cyber threats.

Furthermore, the importance of participating in partnerships has been highlighted, demonstrating how these alliances can lead to more robust cybersecurity measures and policies. Moreover, understanding how to utilize technology for collective defense equips professionals with the skills needed to implement shared security solutions that benefit not their organizations but the broader community.

To summarize, it's crucial to promote a sense of shared responsibility in the realm of cybersecurity, which demands dedication and effort. The various skills acquired in this area, spanning from programs within companies to partnerships and technological advancements, underscore the diverse strategies required to tackle cyber risks effectively. Embracing the idea of responsibility sets the foundation for an online world where collaborative actions result in fortified defenses and a more robust cybersecurity environment.

Summary

This chapter discussed various aspects of effective collaboration, such as establishing trust, upholding confidentiality, utilizing collective intelligence to boost security measures, and encouraging shared responsibility. It began by highlighting the benefits of shared awareness and the collective improvement of capabilities. It also touched upon the skills for participation in threat intelligence platforms and frameworks to take a proactive approach against evolving cyber threats.

Furthermore, it explored the principles of collaboration, underscoring trust and confidentiality as essential foundations. The significance of agreements and secure communication practices was emphasized. This chapter covered skills such as creating and enforcing disclosure agreements, implementing encryption methods, and employing classification frameworks such as the TLP to protect shared data.

Additionally, it demonstrated how actionable intelligence could enhance security operations by improving threat detection, incident response strategies, and vulnerability management. Practical examples were provided to showcase the advantages of incorporating shared intelligence into procedures, from optimizing alert systems to prioritizing patching activities based on real-time threat insights. The skills you acquired included configuring SIEM systems with threat feeds, conducting vulnerability assessments, and participating in incident response initiatives.

This chapter underscored cybersecurity's importance as an effort beyond cybersecurity teams. It stressed the significance of education in promoting a security culture and the advantages of collaborations between private sectors to build robust cyber resilience. The key takeaways from this section involved developing and executing cybersecurity awareness initiatives, participating in partnerships between the private sectors, and using technology to encourage a shared sense of responsibility for cybersecurity.

Index

A

Advanced Persistent Threats
  (APTs) 2, 5, 8, 46, 131
  example 10
aligned cybersecurity initiatives
  impact and value, measuring 107
analyst and third-party testing reports
  external resources, accessing with
    practical examples 130, 131
  findings, applying to organizational
    context 133
  methodologies and results, interpreting 132
  utilizing 130
artificial intelligence (AI) 45
asset inventory
  in cybersecurity 71, 72
  maintaining 73, 74
  updating 73, 74
attacker
  mindset 52
  motivations and objectives 53, 54
  perspective, significance 52
  psychological and behavioral traits 54
  role of mindset, in strengthening
    cybersecurity 55, 56
Attack Surface Management (ASM) 39

attack surface reduction 41
  application security 42
  cloud security 42
  continuous monitoring 42
  incident response planning 41
  inventory and asset management 41
  least-privilege principle 41
  network segmentation 41
  patching and vulnerability management 41
  physical security 42
  regular audits and assessments 41
  secure configuration 42
  system hardening 42
  third-party management 42
  user awareness and training 41
audits 25
Automotive Information Sharing and
  Analysis Center (Auto ISAC) 207

B

baiting tests 31
best practices, for selecting
  security tools 162
  comprehensive market research,
    conducting 163, 164

cost-effectiveness and ROI,
    evaluating 167-169
key stakeholders, involving 164, 165
risk assessment and management 166, 167
big data analytics 60
blockchain-powered security 60
brand reputation 108
Bring Your Own Device (BYOD) 65
business continuity (BC) 103
business impact analysis (BIA) 108-110
business objectives
security measures, aligning with 104
business unit (BU) leaders 105

C

Capture the Flag (CTF) events 55
castle-and-moat approach 77
CCPA
URL 89
Center for Internet Security (CIS) 65, 111
charts 184
chief information security officer (CISO) 86
CIS Controls 69
claims 120
claims, analyzing 125
biases and unsupported assertions,
    identifying 128, 129
contextual analysis, of vendor
    claims 127, 128
skeptical mindset, developing 126
cloud access security brokers (CASBs) 59
**Cloud Security Posture Management
    (CSPM) 55, 73**
cloud workload protection (CWP) 55
collaboration and threat intelligence sharing
best practices, in collaboration
    and sharing 208

imperative, for collaborative defense 206
importance 206
threat intelligence sharing
    mechanisms 207, 208
Colonial Pipeline ransomware attack 17, 18
reference link 18
**Common Vulnerabilities and
    Exposures (CVEs) 26, 112**
comparative visuals 184
**compliance and industry standards, in
    cybersecurity solutions 160**
aligning, with industry norms 161
frequent updates and audits 161, 162
incorporating 161
significance 160
solutions, evaluating for compliance 161
comprehensive asset inventory
building 72, 73
**computer security incident response
    team (CSIRT) 57**
continuous monitoring 43, 80
practices, implementing 74, 75
**Control Objectives for Information and
    Related Technology (COBIT) 65**
**cost-effectiveness and ROI, in
    cybersecurity solutions**
cybersecurity budget, adjusting
    regularly 169
evaluating 167
financial aspects, of cybersecurity 168
financial incentives and grants 169
indirect benefits and cost savings,
    considering 168
negotiations with vendors and
    customization 169
pricing models, evaluating 169
ROI, calculating for cybersecurity
    investments 168

scalability and future expenses of
solution, assessing 168
strategic budgeting 168
total cost of ownership (TCO),
analyzing 168
Cult of the Dead Cow (CDC) 4
customer trust 108
customized cybersecurity strategy
developing 154
cyberattacks, real-world examples 13
Colonial Pipeline ransomware attack 17, 18
NotPetya attack (2017) 13, 14
SolarWinds supply chain attack 14, 15
WannaCry ransomware attack (2017) 16, 17
cybercriminals 4
example 9
cybersecurity
as business imperative 192, 193
asset inventory 71, 72
communication barriers 178, 179
cross-functional teams, building 201, 202
culture, measuring and reinforcing 202, 203
culture of awareness, creating 200, 201
external stakeholders, engaging 198, 199
in business environments 104, 105
interdepartmental collaboration 197, 198
prioritizing, based on business impact 84, 85
role of effective communication 179
shared responsibility, promoting 199, 200
stakeholder roles, defining 196, 197
strategies, for successful collaboration
between technical and non-technical
stakeholders 180, 181
value, communicating to stakeholders 86, 87
**Cybersecurity and Infrastructure
Security Agency (CISA) 207**

cybersecurity-aware organizational culture
importance 192
risks and costs of cyber threats,
assessing 193, 194
role of leadership 194, 195
cybersecurity claims 120-122
cybersecurity landscape 46, 47
cybersecurity market
landscape 163
solutions, comparing 163, 164
cybersecurity measures 50
employee training and security awareness 51
proactive threat assessment 51
regular testing and vulnerability scanning 51
technical defenses implementation 51
valuable lessons, from real-
world incidents 51
cybersecurity mesh 77
cybersecurity solutions
automated tools, using to
improve efficiency 175
continuous improvement, through
feedback and analysis 175
implementing 169, 170
integrating 169, 170
maintaining 174
monitoring 174
respond, to security incidents 175
security audits, carrying out 175
third-party services, significance 175
updating, regularly 174
**cybersecurity solutions, selecting
factors 152, 153**
compliance and industry standard 160
scalability and future-proofing 159, 160
system compatibility and
integration, assessing 154
threat landscape 153

cybersecurity strategy 80
  aligning, with business objectives 83
  comprehensive risk assessments,
    conducting 89-91
  correlation, of organizational goals and
    cybersecurity endeavors 83, 84
  foundational principles and
    components 80, 81
  objectives and goals, setting 81, 82
  prioritization of mitigation strategies 91, 92
  risk management 87
  risk management methodologies, integrating
    in strategy formulation 87-89
  role and significance, of elements 82, 83
cybersecurity tools
  repurposing 143
cyber-terrorists 8
  example 10
Cyber Threat Alliance (CTA) 214
cyber threats 153

D

dashboards 184
data loss prevention (DLP) 144
defense-in-depth 80
Democratic National Committee (DNC) 9
disaster recovery (DR) 108
distributed denial-of-service
  (DDoS) 4, 48, 114, 182
DREAD model 28
Dynamic Application Security
  Testing (DAST) 42

E

Electronic Health Record (EHR) 72
emerging threats
  monitoring, significance 46
  risks 48, 49
endpoint detection and response
  (EDR) 40, 48, 94, 95, 114
endpoint platform protection (EPP) 55
ethical considerations 56
ethical hacking 57
European Union Agency for
  Cybersecurity (ENISA) 207
European Union (EU) 25
extended detection and response
  (XDR) 40, 47, 73, 76, 94
external audits 25

F

Factor Analysis of Information
  Risk (FAIR) 32, 111
FFIEC's guidelines
  URL 89
firewall 76
FS ISAC 207

G

General Data Protection Regulation
  (GDPR) 25, 48, 142, 183, 198
  URL 89
Global Cyber Alliance (GCA) 207, 214
global threat intelligence (GTI) 140, 153
graphs 184

H

hackers-for-hire 9
hacktivism 53
hacktivists 4
  example 10
Hazard Identification, Risk Assessment,
    and Control (HIRAC) 33
Health Insurance Portability and
    Accountability Act (HIPAA) 25, 56, 140
hypothesis-based threat hunting 47

I

identity and access management
    (IAM) tools 145
identity threat detection response
    (ITDR) 55, 95
incident management life cycle 95
  containment and eradication 95
  detection 95
  incident response frameworks 95
  post-incident analysis and review 95
  recovery 95
incident response (IR) 107
  collaboration and communication 97
  critical tools and technologies 96
  human factor 96
  planning and preparedness 92, 93
  role of automation and playbooks 96
incident response plan 75, 76, 92
incident response platform (IRP) 146
Incident Response Teams (IRTs) 76, 96
indicator-of-compromise (IoC)
    based hunting 47
indicators of attack (IoAs) 49
indicators of compromise (IoCs) 207

infographics 184
information security management
    system (ISMS) 88
Information Sharing and Analysis
    Organizations (ISAOs) 210
Information Systems Audit and Control
    Association (ISACA) 65
innovation, in cybersecurity 58
  benefits 58, 59
  driving, within organizations 59, 60
  emerging technologies and future trends 60
insider threats 5
  example 10
intellectual property (IP) 145
interactive data visualizations 184
interdepartmental collaboration,
    cybersecurity
  C-squite executives and
      department heads 198
  finance department 197
  legal departments 197
  operational divisions 197
internal audits 25
International Electrotechnical
    Commission (IEC) 65
International Organization for
    Standardization (ISO) 65
Internet of Things (IoT) 48, 85, 142
Internet service providers (ISPs) 206
Intrusion Detection Systems
    (IDSs) 76, 95, 114, 141
Intrusion Prevention Systems (IPSs) 76, 95
ISO 27001
  URL 88
ISO 31000 32

K

key performance indicators
    (KPIs) 38, 100, 107, 186, 202

L

large language models (LLMs) 60
learning management systems (LMSs) 98
local threat intelligence 153
log analysis 74

M

machine learning (ML) 45
Malware Information Sharing
    Platform (MISP) 207
Mandiant Advantage Threat Intelligence 74
marketing spin 130
mean time to detect (MTTD) 107, 206
mean time to respond (MTTR) 107, 206
memorandum of understanding
    (MOUs) 208
metrics 107
    for security programs and teams 67
MITRE ATT&CK Framework 11, 12, 95
ML and automated threat hunting 47
multi-factor authentication (MFA) 110, 157

N

National Institute of Standards and
    Technology (NIST) 32, 65, 100
National Institute of Standards and
    Technology Risk Management
    Framework (NIST RMF) 110
National Security Agency (NSA) 10

nation-state actors 4
    example 9
NCCIC 216
network detection and response (NDR) 96
NIST Cybersecurity Framework 70, 95, 216
    URL 88
NIST SP 800-30 32
non-disclosure agreements (NDAs) 210
NotPetya attack (2017) 13, 14
    reference link 14

O

Operationally Critical Threat, Asset, and
    Vulnerability Evaluation (OCTAVE) 33
operational resilience 108
organizational vulnerabilities 22-24
    hardware vulnerabilities 23
    network vulnerabilities 23
    real-world examples 24, 25
    software vulnerabilities 23
organizational weaknesses 22
    human 23
    process 22
    real-world examples 24, 25
    technical 22
    techniques, for identifying and assessing 25
OWASP Zed Attack Proxy (ZAP) 217

P

PASTA model 28
Payment Card Industry Data Security
    Standard (PCI DSS) 25
penetration testing 29, 30
Phish Alert button 101
Point of Sale (POS) systems 72, 142, 171
principle of least privilege 80

Private-Sector Offensive Actors (PSOAs) 9
    example 10

Q

quantum computing-based security 60

R

reassessment 42, 43
remediation measures
    implementing 76
request for proposal (RFP) 154
responsible disclosure 57
return on investment
        (ROI) 86, 103, 107, 162
risk assessment 80, 109, 110
    assets, prioritizing 166
    conducting 32
    example template 37
    impact of potential risks, evaluating 166
    in cybersecurity 166
    stakeholders, engaging 167
    strategies, for managing risks 166
    threats and vulnerabilities, analyzing 166
    updating, regularly 167
risk assessment methodologies 32
    Factor Analysis of Information
        Risk (FAIR) 32
    Hazard Identification, Risk Assessment,
        and Control (HIRAC) 33
    ISO 31000 32
    Operationally Critical Threat, Asset, and
        Vulnerability Evaluation (OCTAVE) 33
risk assessment process
    asset classification 33
    asset identification 33
    asset ownership and responsibilities 33

    boundary definition 33
    data flow analysis 33
    data gathering 35
    documentation 35, 37
    impact assessment 35
    likelihood assessment 35
    monitoring and reviewing 38, 39
    regulatory and compliance 33
    reporting 37, 38
    risk scoring 35
    subject matter expertise 35
    weaknesses, prioritizing and remediating 39
risk levels 36, 37, 39
risk matrix 36
risk mitigation strategies 40
    access control measures 40
    backup and disaster recovery plans 41
    cybersecurity awareness
        training programs 40
    data encryption 40
    incident response plans, developing 40
    network segmentation 40
    patch management 40
    secure coding practices, using 41
    security monitoring and logging 40
    third-party risk, assessing and managing 40
role-based access control (RBAC) 145

S

scalability
    in cybersecurity 159
script kiddies 8
    example 10
secure development lifecycle (SDL) 41
Security Assertion Markup
        Language (SAML) 217

security audits 25, 26

security awareness and training
    programs 97, 98

    continuous evaluation and
        improvement 99, 100

    security-first mindset, fostering 101, 102

    tailored training, for organizational
        roles 98, 99

Security Information and Event
    Management (SIEM) 76, 93, 141, 212

security initiatives

    prioritized security initiatives,
        communicating 112

    prioritizing, based on risk and
        business impact 109

    prioritizing, with frameworks 110, 111

security investments

    effective communication strategies,
        developing 115, 116

    technical metrics, translating to
        business value 113-115

    trust, building with stakeholders 117, 118

    value, communicating of 113

security measures

    aligning, with business objectives 104

    and business objectives, connecting 105, 106

security metrics

    implementing 71

    selecting 68, 70

    significance 67, 68

    tracking 71

security metrics selection guidance

    CIS Controls 69

    NIST Cybersecurity Framework 70

security orchestration, automation,
    and response (SOAR) 93

security posture

    components 63, 64

    continuous monitoring 74

    continuous monitoring practices,
        implementing 74, 75

    human factor 66, 67

    improving 74

    technological landscape 76, 77

Security Posture Management (SPM) 39

security processes

    role 65, 66

security programs and teams

    metrics 67

security technologies

    evaluating 64, 65

security tools

    integration 144, 145

    selecting, best practices 162

Server Message Block (SMB) protocol 16

Service-Level Agreements (SLAs) 2

shared responsibility, for cybersecurity

    cybersecurity awareness culture,
        cultivating 215, 216

    promoting 215

    public-private partnerships
        (PPPs), engaging 216

    technology for collective defense 217

shared threat intelligence

    collaborative incident response
        and recovery 213, 214

    integrating, into security operations 212, 213

    leveraging, for improved security 212

small and medium-sized
    enterprises (SMEs) 217

social engineering tests 30, 31

SolarWinds supply chain attack 14, 15

    reference link 16

stakeholders

roles and responsibilities 196

roles, defining in cybersecurity 196, 197

Static Application Security
Testing (SAST) 42

storytelling 182

strategic implementation plan, for
cybersecurity solutions 170

clear goals and objectives, establishing 171

coordination and communication 171

risk management and contingency
planning 172

roadmap, designing for implementation 171

technical and operational
challenges, addressing 171

testing and validation 171

training and support 171

unique requirements, of organization 170

strategies, for collaboration 185

collaborative cybersecurity decision-making
processes, implementing 188, 189

cross-functional cybersecurity
teams, building 186, 187

regular cybersecurity workshops and
training sessions, establishing 187, 188

STRIDE methodology 27, 28

subject matter experts (SMEs) 25

SUNBURST 14

system compatibility and integration
assessment 154

challenges posed by legacy systems,
addressing 155-158

compatibility, evaluating with
existing systems 155

IT environment 155

scaling, for future growth 156

testing and feedback mechanisms,
implementing 158

T

Tactics, Techniques, and Procedures
(TTPs) 1, 10-12, 207

tailgating tests 31

tailored incident response procedures

designing 93, 94

technical concepts, translating for
non-technical stakeholders 181, 182

complex cybersecurity terminology,
simplifying 182

cybersecurity, contextualizing
in business terms 183

effective visualization and presentation,
of cybersecurity data 184, 185

threat actors

Advanced Persistent Threats (APTs) 5, 8

cybercriminals 4

cyber-terrorists 8

hacktivists 4

insider threats 5

motivations 9, 10

nation-state actors 4

objectives 9, 10

Private-Sector Offensive Actors (PSOAs) 9

script kiddies 8

types 1-3

Threat and Vulnerability Management
(TVM) 39, 48, 73, 76

threat-hunting methods

hypothesis-based threat hunting 47

indicator-of-compromise (IoC)
based hunting 47

ML and automated threat hunting 47

TTP- (tactics, techniques, and
procedures) based hunting 47

threat identification 34

threat intelligence (TI) 49, 50, 142

threat landscape 153

constant evolution, of cyber threats 153

customized cybersecurity
    strategy, developing 154

global and local threat intelligence 153

past incidents, assessing 153

threats, unique to different industries 153

threat modeling 27, 29

TI platforms (TIPs) 145

tools

efficiency, maximizing through synergy 145

integration 143-145

repurposing 143, 144

tools and technologies, identifying 139

cybersecurity arsenal, cataloging 140

effectiveness and relevance, assessing 141

future needs, identifying 142, 143

gaps, identifying 142, 143

tool usage optimization 146

advanced configuration and
    customization 146, 147

performance monitoring 147

security audits, conducting 148

training and knowledge sharing 148

Traffic Light Protocol (TLP) 209

trust building 209

among partners 210

confidentiality, maintaining in
    information sharing 210, 211

transparency and confidentiality,
    balancing 211, 212

**TTP- (tactics, techniques, and
    procedures) based hunting 47**

U

Unified Endpoint Management (UEM) 73

**user and entity behavior
    analytics (UEBA) 94**

**user training and adoption, in
    cybersecurity implementation 172**

continuous improvement, through
    feedback loop 173, 174

ongoing refresher courses 173

resistance, addressing to change 173

training effectiveness, evaluating 173

V

vendor claims

contextual analysis 127, 128

facts, separating from marketing 123, 124

substance of cybersecurity solutions,
    evaluating 124, 125

vendors 120

assessing 134, 135

credibility and track record, evaluating 135

customer feedback and post-sale
    support, analyzing 136, 137

offerings, aligning with organizational
    requirements 137

virtual private networks (VPNs) 153

visual aids 182

vulnerability assessments 26, 27, 34

W

**WannaCry ransomware attack
    (2017) 16, 17, 141**

reference link 17

web application firewall (WAF) 147

www.packtpub.com

Subscribe to our online digital library for full access to over 7,000 books and videos, as well as industry leading tools to help you plan your personal development and advance your career. For more information, please visit our website.

Why subscribe?

- Spend less time learning and more time coding with practical eBooks and Videos from over 4,000 industry professionals

- Improve your learning with Skill Plans built especially for you

- Get a free eBook or video every month

- Fully searchable for easy access to vital information

- Copy and paste, print, and bookmark content

Did you know that Packt offers eBook versions of every book published, with PDF and ePub files available? You can upgrade to the eBook version at packtpub.com and as a print book customer, you are entitled to a discount on the eBook copy. Get in touch with us at customercare@packtpub.com for more details.

At www.packtpub.com, you can also read a collection of free technical articles, sign up for a range of free newsletters, and receive exclusive discounts and offers on Packt books and eBooks.

Other Books You May Enjoy

If you enjoyed this book, you may be interested in these other books by Packt:

Okta Administration Up and Running

HenkJan de Vries | Lovisa Stenbäcken Stjernlöf

ISBN: 978-1-83763-745-4

- Get a clear overview of Okta's advanced features

- Integrate Okta with directories and applications using hands-on instructions

- Get practical recommendations on managing policies for SSO, MFA, and lifecycle management

- Discover how to manage groups and group rules for Joiner, Mover, Leaver events in Okta using examples

- Manage your Okta tenants using APIs and oversee API access with Okta

- Set up and manage your organization's Okta environment, ensuring a secure IAM practice

- Find out how to extend your Okta experience with Workflows and ASA

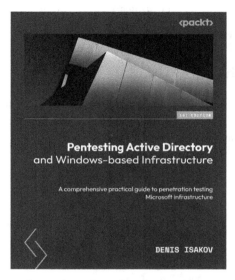

Pentesting Active Directory and Windows-based Infrastructure

Denis Isakov

ISBN: 978-1-80461-136-4

- Understand and adopt the Microsoft infrastructure kill chain methodology
- Attack Windows services, such as Active Directory, Exchange, WSUS, SCCM, AD CS, and SQL Server
- Disappear from the defender's eyesight by tampering with defensive capabilities
- Upskill yourself in offensive OpSec to stay under the radar
- Find out how to detect adversary activities in your Windows environment
- Get to grips with the steps needed to remediate misconfigurations
- Prepare yourself for real-life scenarios by getting hands-on experience with exercises

Packt is searching for authors like you

If you're interested in becoming an author for Packt, please visit authors.packtpub.com and apply today. We have worked with thousands of developers and tech professionals, just like you, to help them share their insight with the global tech community. You can make a general application, apply for a specific hot topic that we are recruiting an author for, or submit your own idea.

Share Your Thoughts

Now you've finished *Cybersecurity Strategies and Best Practices*, we'd love to hear your thoughts! Scan the QR code below to go straight to the Amazon review page for this book and share your feedback or leave a review on the site that you purchased it from.

https://packt.link/r/1803230053

Your review is important to us and the tech community and will help us make sure we're delivering excellent quality content.

Download a free PDF copy of this book

Thanks for purchasing this book!

Do you like to read on the go but are unable to carry your print books everywhere?

Is your eBook purchase not compatible with the device of your choice?

Don't worry, now with every Packt book you get a DRM-free PDF version of that book at no cost.

Read anywhere, any place, on any device. Search, copy, and paste code from your favorite technical books directly into your application.

The perks don't stop there, you can get exclusive access to discounts, newsletters, and great free content in your inbox daily

Follow these simple steps to get the benefits:

1. Scan the QR code or visit the link below

https://packt.link/free-ebook/9781803230054

2. Submit your proof of purchase
3. That's it! We'll send your free PDF and other benefits to your email directly

www.ingramcontent.com/pod-product-compliance
Lightning Source LLC
Chambersburg PA
CBHW080636060326
40690CB00021B/4959